W0036041

SAGE was founded in 1965 by Sara Miller McCune to support the dissemination of usable knowledge by publishing innovative and high-quality research and teaching content. Today, we publish over 900 journals, including those of more than 400 learned societies, more than 800 new books per year, and a growing range of library products including archives, data, case studies, reports, and video. SAGE remains majority-owned by our founder, and after Sara's lifetime will become owned by a charitable trust that secures our continued independence.

Los Angeles | London | New Delhi | Singapore | Washington DC | Melbourne

How a 24X7 Digital Marketplace Is Transforming Business

ROUND THE CLOCK

How a 24X7 Digital Marketplace Is Transforming Business

ROUND THE CLOCK

RAY TITUS

Los Angeles | London | New Delhi
Singapore | Washington DC | Melbourne

First published in 2019 by

SAGE Publications India Pvt Ltd
B1/I-1 Mohan Cooperative Industrial Area
Mathura Road, New Delhi 110 044, India
www.sagepub.in

SAGE Publications Inc
2455 Teller Road
Thousand Oaks, California 91320, USA

SAGE Publications Ltd
1 Oliver's Yard, 55 City Road
London EC1Y 1SP, United Kingdom

SAGE Publications Asia-Pacific Pte Ltd
18 Cross Street #10-10/11/12
China Square Central
Singapore 048423

Published by Vivek Mehra for SAGE Publications India Pvt Ltd, typeset in 11/14 pts Minion Pro by Fidus Design Pvt Ltd, Chandigarh and printed at Chaman Enterprises, New Delhi.

Library of Congress Cataloging-in-Publication Data

Name: Titus, Ray, author.
Title: Round the clock: how a 24×7 digital marketplace is transforming business/Ray Titus, Professor of Marketing & Dean, Alliance School of Business, Consulting Advisor.
Description: New Delhi, India; Thousand Oaks, California: SAGE Publications India, 2018. | Includes bibliographical references.
Identifiers: LCCN 2018041617 | ISBN 9789352808205 (pbk) | ISBN 9789352808229 (web)
Subjects: LCSH: Information technology—Economic aspects. | Technological innovations—Economic aspects. | Electronic commerce. | Internet marketing.
Classification: LCC HC79.I55 T58 2018 | DDC 381/.142—dc23
LC record available at https://lccn.loc.gov/2018041617

ISBN: 978-93-528-0820-5 (PB)

SAGE Team: Manisha Mathews, Syeda Aina Rahat Ali and Ankit Verma

To
Amma & Appachen

Thank you for choosing a SAGE product!
If you have any comment, observation or feedback,
I would like to personally hear from you.

Please write to me at **contactceo@sagepub.in**

Vivek Mehra, Managing Director and CEO, SAGE India.

Bulk Sales

SAGE India offers special discounts
for purchase of books in bulk.
We also make available special imprints
and excerpts from our books on demand.

For orders and enquiries, write to us at

Marketing Department
SAGE Publications India Pvt Ltd
B1/I-1, Mohan Cooperative Industrial Area
Mathura Road, Post Bag 7
New Delhi 110044, India

E-mail us at **marketing@sagepub.in**

Subscribe to our mailing list
Write to **marketing@sagepub.in**

This book is also available as an e-book.

Contents

Preface

More than two decades ago, completing my undergrad programme in hospitality, I joined one of India's leading hospitality hotel chains, the Taj Group of Hotels. Being a food and beverage (F&B) service specialist, I worked at the restaurants and other service outlets in Taj's hotel in Bangalore. It was a tough job with long shifts and working weekends. As an F&B service professional, I was required to be well versed with menu at the various outlets in the hotel. Every few years when the chefs decided to change the menus, there would be a flurry of activity at all the F&B outlets. The new dishes on the menu would be carefully designed after weeks of deliberations and then tested before they became part of the 'official menu'. The testing included service professionals and even some regular customers sampling the new dishes and participating in finalizing on the ones that finally got on to the menu card. The whole exercise would last a month or two before things settled down to the routine of a new menu. The exercise of changing the menus came around almost every year.

Here's a stark contrast to what I experienced while at the Taj. In the world of digital, menus change every day! Or at least, an everyday changing digital menu becomes part of a firm's differentiated proposition for the consumer. FreshMenu, a food technology start-up that operates on a cloud kitchen food-delivery model where it churns out its own signature dishes and then delivers, proclaims thus on their portal, 'Our menu changes daily because we don't believe in boring you (or ourselves). We also like to keep you guessing. Cliff-hangers, that's what we believe in. But here's a spoiler. The menu changes, the goodness doesn't'. FreshMenu, seen as a full stack online kitchen, was founded in 2014, and first started out of Bengaluru, the city I live in. It's now moved into two other metropolitan cities of India, Delhi and Mumbai.

FreshMenu does every day what the Taj F&B outlets do every year. Imagine that. The one thing that's made FreshMenu possible is digital. It's the digital avatar FreshMenu adorns that allows it to keep its menu changing every day. It's again digital that allows FreshMenu the incredible flexibility in altering prices, offering discounts, running promos and targeted campaigns, and managing every other possible piece that makes up the value proposition it presents to its online buyers. What's more, the online outfit aided by digital can even customize the value it offers down to a single buyer who patronizes the brand online.

The enormous power of digital technologies and platforms, I believe, is an all-round game changer—that means every sphere that pervades human living. As I write this, the #MeToo campaign is exploding around me. With zero marketing spends, this campaign has exploded around the world. What digital has done for #MeToo is beyond just spreading the word among the digital crowd. Until now, digital platforms have been described as propagators and aggregators. I am saying that's just the tip of the iceberg. As in the case of #MeToo, digital enables communities to band together and seize genuine power. Such empowerment is an outcome of a transparent-inclusive action that leaves out no one, at least not anyone who wants to join in. Imagine the credibility such groups carry! Their messages have the power to persuade—persuade to the extent of genuine change!

The power of digital is everything business and marketers should have hoped for. Of course, for the old-timers in the business who have enjoyed the until now 'command and control' model, the 'one to many' propagation platforms may not be cheering, but the genuine ones seeking pioneering ways to create, communicate, deliver and service superior value to customers should be overjoyed. To them I say, your starting point is the digital ecosystem that houses a digital marketplace. That is where everything is rapidly and radically changing. Embrace this marketplace for what it allows buyers and sellers to achieve. In fact, it's precisely for this reason I have written *Round the Clock*. It's to enable you to open your businesses to, and seize the tremendous value-creating opportunities digital markets hold. It's to enable you to navigate a marketplace that is nothing like you've been used to in the past. Note that all the changes you see in digital selling and buying and everything else associated with it are neither the seller nor the buyer's doing. It's the digital marketplace that's changed them all. So when an OYO comes along, travellers who use the platform to book hotel rooms in addition to finding the best deal get an opportunity to eliminate the 'unpredictability problem' that's been plaguing them. The unpredictability problem refers to zero guarantees on how a product and service will eventually turn out, once the digital buyer makes a purchase. OYO plugged that problem by working with their hospitality partners and ensuring the OYO brand delivered on a promise of good hospitality. Both the aggregation of available hospitality spaces and the guarantee of a quality service got going, thanks to digital. It's digital technology that connected OYO to a traveller and it's again digital that allowed for buyers to believe in the OYO promise. Now let's, for a moment, assume that OYO didn't deliver. It will again be digitally connected travellers who will warn each other off from patronizing the brand. Of course, if OYO delivered, their happy guests can tom-tom the digitized good news to others in no time.

The transformational impact of digital technologies, applications and platforms is leaving no marketplace untouched. Almost every time when I walk into a classroom to teach a course, I am reminded of how none of what I do now may last. Experts around the globe have identified various digital transformation trends in teaching and learning that include augmented, mixed and virtual reality in classrooms. An app like Unimersiv can transport students to virtual places like the international space station or the Acropolis of Athens, or even Dino Land. The ubiquitous availability of smartphones and low-priced virtual reality (VR) headsets and their use in classrooms can ensure that a student's cerebral cortex is tricked into experiencing a chase by a scary Tyrannosaurus rex. On a personal note, in my strategy classes when I talk of how Alexander beat Porus in the Battle of the Hydaspes by the banks of river Jhelum, my students can actually be in the thick of the battle via their VR headsets. Will that have an impact on their absorptive capacity? Studies that measured biometric eye tracking, electrodermal response and heart rate changes revealed that users who were plugged into virtual reality had a 27 per cent greater emotional engagement with simulated content than when it was presented in two-dimensional traditional videos. Frankly, that excites me as an educator.

In the end, *Round the Clock* is about unravelling the digital marketplace by focusing on the three elements that make it all up, namely the virtual marketplace, virtual sellers and virtual buyers. The book is your chance to sign up to this brave new digital world and reap its benefits.

I can assure you that if you manage your digital value proposition well enough to drive better value to your target buyers and keep at it consistently in a rapidly altering marketplace, your pay-offs are more than assured!

Wishing you the very best. Cheers.

Acknowledgements

Writing *Round the Clock* has been quite an endeavour in learning for me as a marketing academic and professional. The world of digital has been evolving at a pace so quick that research has found it difficult to track and document its progress. That explains why research material on the digital marketplace, for example, are few and far in between. In trying to put this book together, I have dipped into the limited studies that I have done while leaning heavily on research work done around the world. I am, thus, thankful to all those knowledge creators whose work has helped me present my propositions in the book. Writing about an arena that is constantly and rapidly evolving has been both challenging and exciting.

I am grateful to have found the support and time at my place of work for my writing. Teaching at the Alliance School of Business for more than a decade and a half has been rewarding in every sense. The opportunity to work with colleagues I have, and that include Dr Debashish Sengupta, has been intellectually enriching to say the least. I have also learnt much from my students in the classroom. Even after they leave campus, the bond I share with them is a testimony to what shared learning can achieve.

The team at SAGE has been wonderful. I am deeply thankful to my Commissioning Editor Manisha Mathews for her constant support and encouragement. Her patience and faith in the book made things easier for me as I toiled at my writing.

When it comes to family, their love and support has been unconditional as always. Knowing that Appachen and Amma are always there for us is truly comforting. My brother Thomas, his wife Ruby and their lovely girls, Abigail, Amelia and Julianne, have always been a bedrock of love and care. Having Appa, Ma, Mathew and Celeena as part of our support system has been comforting. My wife Alphy is an epitome of grace and dignity, always reminding me of the better person I can be. She is the class act I aspire to be. My kids, Jaden and Brooklyn, have taught me what it is like to find happiness in the littlest of things. For that I am grateful. The love we share as a family fixes me in my head like nothing else can. Thank you, guys.

Finally, I am humbled by the eternal love I found at the Cross of Calvary. Knowing that I have been saved by grace is the greatest of all knowledge. In everything, therefore, I say, let thy will be done.

Amen.

CHAPTER 1

Shifting Paradigms

While discussing a recent research assignment with executives at a consumer electronics major headquartered in Bengaluru, I was encouraged by the tone and tenor of what I was hearing. The marketing people at the company were talking about launching a mobile application (app) that would allow users to control their TV-viewing experience and customize content they wanted to consume. The company which sold smart TVs wanted to revolutionize their buyers' viewing experience. Intrigued, I asked them why? The answer I got is a pointer to the business mindset necessary for the world of digital.

> You see, we believe we are in the business of building and customising entertainment solutions for our customers. In keeping one step ahead of our competitors, we build and use inhouse technologies to ensure from the time our customers buy and put our TVs on, they are in control of their viewing experience. We don't want broadcasters or TV hardware to dictate to our viewers how they access and consume content. That's why we are launching this app. We want our users to move to a 'smart' remote control that can enable them to dictate their viewing experience. Our App allows users to customise content and access. Now this is transformational in that we are using digital applications to put our buyers in charge. What we intend to do is to co-create a viewing experience that is customised to a single user!

I nodded in agreement. Smart move, I thought to myself. In an era of digital enablement and transparency, the right side to be on is where selling–buying experiences are co-created by businesses and their customers.

Seen from a holistic perspective, digital is ultimately about a connected space that puts people, devices and objects from everywhere within each other's reach via a virtual platform. For the last half a decade, I have been studying the power of the digital medium and enabling technologies. Although I started with a focus on the user (I teach consumer behaviour, you see), over time I've stepped on to either sides to see how digital technologies are enabling value creation for both buyers and sellers. Based on my research and consulting work with businesses that have unlocked and leveraged the value embedded in digital, I have come to conclude that the era of digitization is one that's a win-win—for businesses and consumers. Most people seem to think that this is an era that's firmly siding with digital consumers. Surely it is. However, that's only one side of this boom story. This is also the era of digital businesses. Data show companies that have embraced and leveraged the power of digital have been able to transform their customers' experiences, and engage with their target buyers for mutual benefit. If today buyers use virtual information to make smart buying decisions in a digital marketplace by picking and choosing the best solutions at the best prices possible, businesses in response are tapping into a ton of information about their buyers, their preferences and the marketplace, of course, at times, even to the extent of being intrusive. Sellers leveraging digital are able to track buyer journeys to the minutest of detail, and accordingly they have designed and custom-built customer engagements that have ended in both a sale and a digital lock-in for probable lifetime value.

The digital business playground is one that is open and transparent both ways. Until now, our understanding of a market has been limited to a space where buyers and sellers engage and transact. That is no longer relevant and is a dangerously limited reading of a marketplace. Today's digital markets are 'borderless' and 'boundaryless'. Borderless in that buyers and sellers aren't restricted by geographies, and boundaryless in that digital marketplaces aren't time-bound and stretch on until an infinite period of time. These are reasons why I have termed the digital marketplace as an 'infinite market'. Digital market engagements begin far ahead of an actual transaction. They begin when the buyer steps on to the virtual marketplace. Now such footprints may or may not land on seller radars. But those footprints remain there in perpetuity, available for 'tracking'. Also, even when the digital buyer's journey ends with a purchase and feedback, it really isn't over. Truth is that in a digital marketplace, there is no end to a customer's journey and the impact it can have on other potential buyers. When a digital buyer uses digital media platforms to propagate information about his/her buying and consumption experience to other prospective buyers, he/she does put it out there in perpetuity.

Let me quickly illustrate this with a digital information search a colleague did a couple of months ago, and the impact it had on his decision to buy. He had recently recovered from a throat infection. The unpredictable weather changes in Bengaluru did him in, so he decided to consult an ENT specialist. His decision on which hospital and doctor to go to was done on the Practo app he had loaded on his phone. Practo, a Bengaluru-based health care service aggregator, positions itself as 'your home for health'. A quick search for hospitals with an ENT close to where my colleague lived threw up close to 20 options. Once he had filtered the 20 to a set of 5, he started his evaluation. Soon he was ready to book an appointment with a leading multispeciality hospital close to where he lived. Then something happened. On closer evaluation of his choice, he found that there was a negative review given by a former patient to one of the doctors at the hospital he had chosen. A quick reading revealed how this doctor had botched up on the diagnosis he had made. The review was scathing and the patient in question recommended that no one else should visit this particular hospital. If they did, they would be doing so at the risk of a misdiagnosis. Even though the review was from three years ago, my colleague told me that it was enough to sow seeds of doubt about the hospital in his mind. He was now having second thoughts about going to this hospital, never mind it being a branded super speciality health care centre. He made up his mind and decided that he didn't want to go to a hospital where a doctor got the diagnosis wrong.

Here's how my colleague finally made his decision. After a thorough reading of posted reviews on the app, he ended up visiting a private clinic situated close to the multispecialty hospital. What about his throat? 'Fit as a fiddle', he told me. Did it end with that? No. My colleague, a week after he recovered, got back on to the app and wrote a glowing review about the clinic he had visited and the doctor who treated him.

In a study[1] from a few years ago on global shopper attitudes towards buying consumable products online, it was found that 7 out of 10 buyers liked to read reviews prior to making a purchase decision. A substantial 61 per cent of digital buyers spent considerable time researching products online before making up their mind. Almost a half of them used social media sites to aid them in their buying choices. Out of 10 digital buyers, 4 make up their mind[2] by reading just one to three reviews. From a consumer persuasion perspective, this is ample proof that the power to influence has firmly shifted from marketer content targeted at buyer groups to free-floating digital content created and propagated by fellow consumers. Of course, it is easy to see why the latter has greater powers to persuade buyers. With marketer trust hitting rock bottom levels, online reviews stand up for higher credibility in the eyes of potential buyers. Note that such reviews aren't just purchase persuaders. Glassdoor specializes in employee reviews that aid potential job seekers in making their minds up about organizations

[1] Nielsen, *Ecommerce: Evolution or Revolution in the Fast-moving Consumer Goods World* (Nielsen Global E-commerce Report, August 2014). Available at: http://www.nielsen.com/content/dam/nielsenglobal/apac/docs/reports/2014/Nielsen-Global-E-commerce-Report-August-2014.pdf (accessed on 17 July 2018).

[2] Khusbu Shrestha, '50 Important Stats You Need to Know About Online Reviews', Vendasta Blog, 21 May 2018. Available at: https://www.vendasta.com/blog/50-stats-you-need-to-know-about-online-reviews (accessed on 17 July 2018).

they want to work with. RateMyProfessors puts out information about university academics by students who have taken classes, which in turn assists college-goers in picking course classes and professors.

I want to reiterate that digital media content in the form of reviews should not be seen only as a 'threat'. In fact, the negative review my colleague read about a doctor at a super speciality hospital may have cost the health care provider a patient, but it did win a private clinic a new patient. Plus the review my colleague put up about the clinic and the doctor who serves there will stay out there in the digital ecosystem for years to come, persuading other buyer-patients to try the clinic's services out. Keeping in mind the potential digital markets' offer to both sellers and buyers, I have structured and sequenced my book into three distinct sections. There is a legitimate logic to this sequence which I will explain after describing the three sections.

The first section of the book focuses on the digital marketplace. This market is nothing like what businesses and buyers have seen in the past. If there is a market that can be truly called 'hyper-dynamic', this is the one. Constantly evolving, digispace's stability is continuously punctured and disrupted by changing technology, or at least in the innovative ways digital technology is being leveraged to create superior value propositions. Such innovative usage and disruptions aided by technology bring to the surface unique ways and models of doing business. Let me illustrate this with an example from retail. Everlane, an online-only clothing retailer headquartered out of San Francisco, California, built and propagated its digital retail identity on the core tenet of what they called 'radical transparency'. In a business where buyers shopping for clothes probably wonder about how much money brands were making on their price tags, Everlane did an about turn and broke the retail code—that of costs and margins to company, and price to shoppers. They pulled away the curtain on masking costs and margins in the apparel business by telling their shoppers that they had a right to know why they were being charged the prices on the tags. Positioning-wise, Everlane enacted a retail branding coup by hooting up the 'transparency tree'. It's no wonder why they say this up front to apparel shoppers[3]:

Your designer clothes sell for 8 times what they cost to make.

Not at Everlane.

We believe customers have the right to know where their products come from and what they cost to make.

We started Everlane five years ago with a simple statement:

Your designer clothes sell for 8 times what they cost to make. Not at Everlane.

Those words struck a chord.

We set out to design beautiful basics, and offer them online only without any of the traditional markups.

That means a luxury tee that costs $8 to make is sold to you for $15—instead of $50.

We soon realized—not only did people not know the crazy markups they pay, they didn't know where their clothes were being made, or under what conditions.

The idea of Radical Transparency quickly became a core tenet of Everlane, influencing decisions at every level of our business. We now reveal all of our costs and document every factory we work with.

Meet a new kind of retail.

---

[3] https://www.everlane.com/about (accessed on 11 September 2018).

Everlane calls itself 'disruptive' in the way it sells clothes. The pivot of 'radical transparency' that Everlane uses to position and differentiate itself works because it has leveraged the digital medium to put across a 'radical' business approach that gets buyers to stop and think about their apparel purchases. Let me explain this in behavioural terms from the perspective of a buyer. Most of the premium and luxury fashion brands out there use the power of imagery and PR to craft an identity of 'upper class'. They then back that up by charging from buyers prices that are way higher than those charged by budget brands. Shoppers on their part agree to the transaction as they believe they are buying 'status value'. Now such a purchase is an emotive one. There is hardly any cognitive utility-based information processing a shopper engages in before a luxury brand purchase. Enter Everlane. Using the power of a high-involvement medium that is digital, they put out the 'real story' behind those inflated premium/luxury brand prices. A shopper engaging with Everlane on the digital platform takes time to read through the online retailer's 'radical transparency' proposition. In doing so, the buyer steps into a cognitive information-processing zone. The buyer is now moving away from an emotive territory to a rational one. Such thinking by the shoppers is good for Everlane, for it allows them to construct a buying value proposition that makes 'rational sense' to shoppers. As proof of its business claim, Everlane puts up a tee on its webpage and illustrates how, based on costs, a tee it retails at US$15 online is sold by the big brand at an 'unfair' high of US$50.

Remember that buyers on digital mediums engage with content at their pace. That is why this medium is characterized as a self-paced medium. When buyers have the time to take in information that businesses like Everlane put out, they process it cognitively to make relatively more rational buying decisions. Right now, when I visit the Everlane portal, I am greeted by 'The Form Slide Sandal'. What is the price Everlane is charging a buyer for the sandals? US$98. What is Everlane telling the buyer about the price other retailers are charging for the same sandal? US$225.

For Everlane, that is probably a deal done!

Sure, 'radical transparency' is the differentiating feature of Everlane, but what allows for the concept's execution in a marketplace is the power of digital. Everlane, according to its CEO, Michael Preysman, operates on a limited marketing budget that's dedicated to digital and social. It's grown on since its launch via recommendations and word-of-mouth spread. You see, the web shopper who is engaged via the digital medium, who then turns into a buyer further ahead, dons the role of a recommender. All this happens because the information processing that digital allows for ensures that buyers are locked in with a cognitive logic that appeals to their rational buying senses. These buyers, in turn, pass on that logic via the same medium to other shoppers, ensuring a link is formed and spreads fast. Now I agree that digital is not a guarantee for either a purchase or loyalty; however, it allows for businesses to engage and propagate in a manner no medium does.

The study of the digital medium, tools and apps that make up the rapidly evolving digital marketplace is what the first section of the book focuses on.

The following section moves into digital buyer territory, aka into the world of digital consumption decision-making. Understanding buyers necessitates an understanding of three elements that make up a buyer's decision-making journey. The first of these is the sociocultural environment that houses agents of influence, namely marketers and reference groups. Both these entities are agents of influence. The second element to a buying journey is the psyche that buyers bring into a purchase play, and the final piece of the puzzle is the actual decision-making process. It's pertinent to ask if digital has invaded the buyer's journey and infiltrated the three elements listed as part of that journey. The answer is, you bet! First, digital has altered the influencer environment. It's firmly taken away the power to influence from the hands of marketers and put it into the hands of a digital public who are now calling the shots. Now don't get me wrong. I am not saying that digital has neutered marketer influence. I am only pointing to how traditional media marketing is running its course. Investments that are being made on marketing content propagation via traditional media by brands are losing steam. It's finding no resonance with buying groups. In fact, it's even being subject to public scrutiny on digital media by various stakeholders, and such transparency is hurting brands.

Second, digital content is influencing buyer psyches like never before. Such content is capturing buyer attention, forming perceptions, engineering learning and dictating attitudes. Furthermore, digital brand personas are engaging with consumer groups to form bonds of loyalty. Finally, every aspect of the buying decision journey, namely problem recognition, information search, evaluation of alternatives, purchase and post-purchase behaviour, is being facilitated by digital tools, technologies and platforms. In effect, it's now valid to say that buying and everything to do with it have gone comprehensively digital! In the current times of digital commerce and surely in the future, organic user-created content on digital mediums will influence buyer psyches and prompt brand adoption. Traditional marketer messaging will be discarded in favour of user-propagated content. The latter, seen as far more credible stimuli, will induce buyer perceptions. If managed well, it will further drive buyer learning and attitude formation. Such digital media-driven attitudes will in the end dictate buying behaviour. In fact, marketers will first gingerly add digital blocks into their worldly business existence and then, over time, will embrace a digital avatar wholeheartedly. When I say this, I am not just referring to the marketing function alone. Note that for businesses, this transition is not a matter of choice. It is a necessity for survival. In the business of technology, this switchover is almost complete. There are even firms out there who define their existence not based on the technology products they sell but on their intangible virtual make-up.

It's no wonder then that Lei Jun, co-founder of China's Apple, Xiaomi, calls his mobile phone firm an e-commerce company. 'We're actually an Internet company. We've already got a business in mobile phone hardware and we want to add to that an Internet platform. We can earn money from that, once it's established'. When Reuters' Jane Lanhee Lee pressed[4] Lei with how his firm could call itself an Internet company when it sells mobile phones, the reply was revealing:

> Well, I'd like to turn the question around and ask, what is an Internet company? The Internet represents an advanced way of thinking. Companies that are armed with this have an incredible competitive edge. Today, we're already China's third largest e-commerce company and we've already built a massive, massive, mobile internet platform. And we have a lot of apps. We've built the MIUI firmware based on the Android OS and it's a very strong system. People just don't get it. The mobile phone itself is only the carrier. Microsoft used to sell Windows in a box with a CD in it. Does that make Microsoft a paper box company? The box and the CD are only the carrier. If people don't understand this, they can't understand what kind of company Xiaomi is.

Again, Lei responded to a comparison of Xiaomi to Apple with this:

> I think the core problem is people don't understand Xiaomi. The main difference is that it is the internet that has made Xiaomi. Apple's design is very simple, and no matter who the customer is, there is one set solution. There are few opportunities to meet individual preferences. But Xiaomi's system can be

---

[4] Henry Blodget, 'CEO of "China's Apple" Is Insulted by Comparison to Apple—Says They're More Like Google or Amazon', Business Insider, 15 August 2013. Available at: http://www.businessinsider.com/ceo-of-chinas-apple-xiaomi-lei-jun-2013-8?IR=T& (accessed on 17 July 2018).

changed by anyone, and thousands of designers create all sorts of functions, looks and solutions.

The all-round practice of digital in every walk of Xiaomi's existence is truly remarkable. It is their digital outreach that's allowed Xiaomi to grow its cult following and engineer rapid adoption of its products among its target buyers. Its use of Weibo, a microblogging site that's wildly popular in China, is an example of the way Xiaomi uses digital social platforms to connect with its adoring fans. 'Mi Fan', as they are called, exhibit cult-like devotion to Xiaomi. They devotedly turn up in large numbers for the brand's annual Mi Fan Festival. Now in the digital world, it's difficult to keep up the good show for long. Even Xiaomi faces its hurdles. With brands like Oppo and Vivo upping their marketing spends, Xiaomi will have to innovate and move at a rapid pace to stem its drop in market share. The way out will be through the minds of finicky digital buyers who are bombarded with digital content 24 × 7 which they access in real time via their digital devices.

The final section of the book does a deep dive into business firms that have ensconced digital firmly into their marketing and other functions. I also do a round-up of enabling firms that are helping businesses leverage digital data and analytics to take smarter business decisions. Take the business of bricks-and-mortar retail for example. Conventional wisdom seems to suggest that bricks-and-mortar retail has limited opportunities when it comes to digital. Not so at all. It is, in fact, the opposite. Digital opportunities abound in the physical space but go abegging most times.

When Gary Angel, founder and CEO at US-based Digital Mortar, launched his firm, his mission was to leverage customer journey analytics to power the reinvention of physical retail. You see, bricks-and-mortar retail businesses the world over have been slow to use digital technology, tools and apps to capture in-store shopper insights and use them to enhance customer experience. Now that's a hugely wasted opportunity that is costing physical retail and stagnating revenues. With digital commerce sniping at their heels, the only feature bricks and mortars can truly use to get customers into their stores is by ensuring that when shoppers get in, they have an enthralling store 'experience'. Digital Mortar on its part intends to track, capture and bring digital buyer behaviour into a retail decision-making zone. The firm specializes in providing cutting-edge measurement and analytics tools for optimizing physical spaces. They are adept at tracking in-store customer journeys and aiding in optimizing store layouts, merchandising and staff performance. It's important to note Gary's ominous prophetic words[5] about the future of retail:

> So if you're feverishly building new stores, designing new store experiences, buying into cutting edge digital integrations, or betting the farm on new uses for your real-estate, wouldn't it be nice to have a way to tell if what you're trying is actually working? And a way to make it work better since getting these innovative, complex things right the first time isn't going to happen? This is the bottom line: these days in retail, nobody needs to invest in customer measurement. After all, there's a perfectly good alternative that just takes a little bit longer. It's called natural selection. And the answers it gives are depressingly final.

The writing is squarely in the air. The opportunities digital presents to business and buyers are enormous. My endeavour with this book is to put the pieces of the digital puzzle together, that of the marketplace, the seller and the buyer. I am hoping that by the time you are done with the book, you are able to see digital for the opportunities it presents. I firmly believe, and hopefully you

---

5 Gary Angel, 'Change or Die: Lessons from the Retail Apocalypse', Digital Mortar Blog, 4 May 2017. Available at: http://digitalmortar.com/change-die-lessons-retail-apocalypse/ (accessed on 17 July 2018).

will too, that digital represents the best chance we have as sellers and buyers to co-create our experiences. It is not good to see our digital footprints giving away information about ourselves to the other party. To the contrary, businesses and buyers now have a chance to tell each other about their respective lives. They now have the opportunity to partner each other, so there's commercial value flowing both ways. The adversarial relationship that is common to the act of buying and selling can be replaced by relationship engagements that deliver value to either parties. You see, that's the promise of digital.

A note of caution: I may not be able to capture everything that makes up the digital world. Neither will I be completely up to date on the disruptive progress that pushes digital practices even further. However, my goal of letting you experience what digital does and may do for the future will, I believe, be valuable learning that you can carry into the world of business and commerce.

Here's hoping you do that.

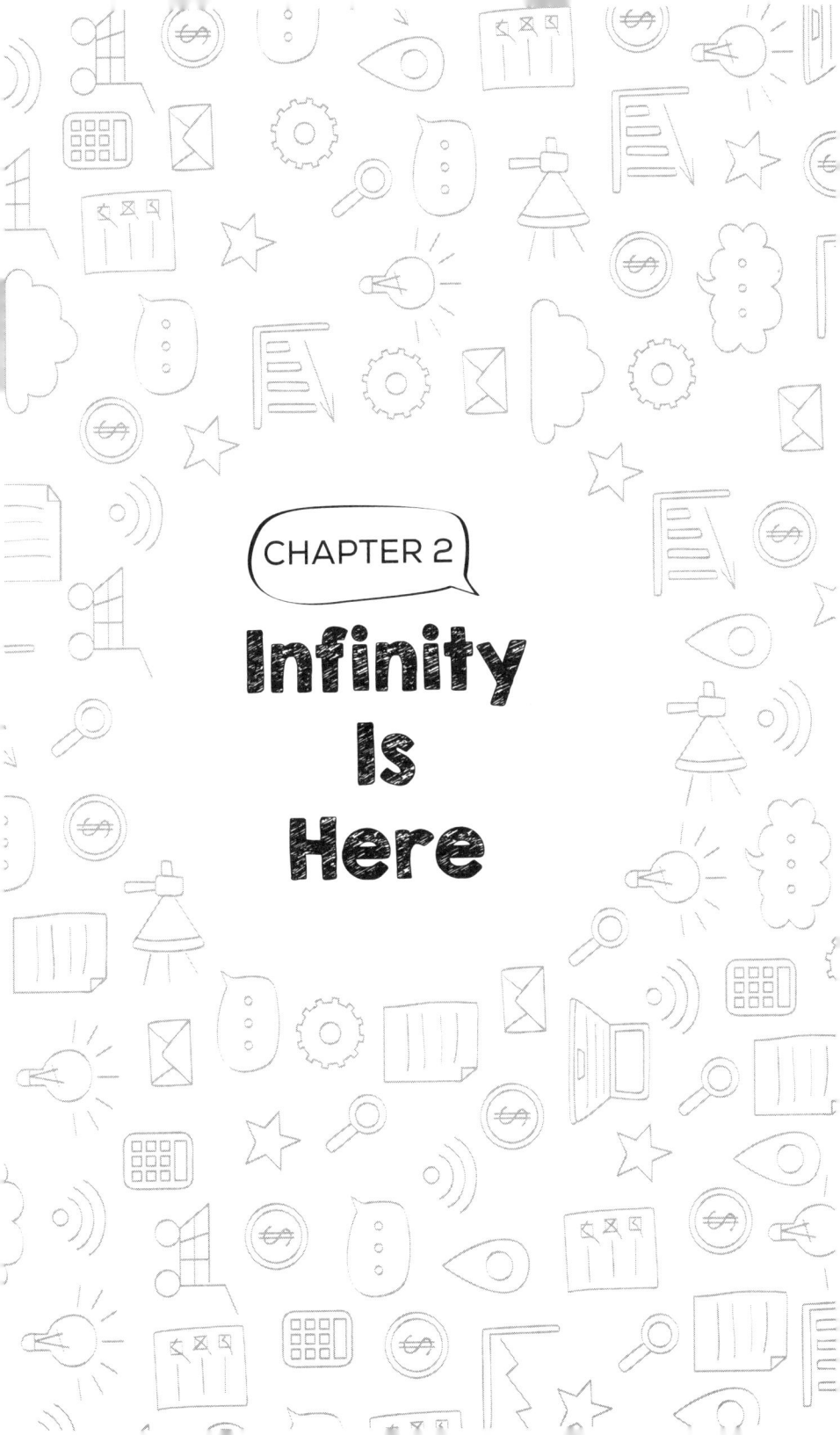

CHAPTER 2

Infinity Is Here

When my colleagues suggested we go to the Jama Masjid area in Delhi for some shopping they wanted to do, I agreed without any hesitation. I was in the capital city with regard to work. I found myself free for the evening and thought what a better way could be than to spend time watching real buying and selling in the crowded marketplaces that dotted the city of Delhi. Of course, I was also looking forward to seeing if the digital stuff mattered in places such as Jama Masjid, Karol Bagh, Chandni Chowk and other such markets. I was convinced it did, but I had to see the evidence myself. Boarding the Delhi Metro, I looked around to see if the people on board were busy on their personal devices; sure enough, many of them were. Although the percentage of digitally inclined favours the young, I spied that a few from the older generation were glued to their mobile phones. The data for India comprehensively confirms that digital adoption is on its upswing in the country. Today, almost a billion Indians own mobile phones. An equally impressive number of 300+ million Indians have Internet access and that number is set to hit a mammoth 800+ million in the next 10 years. Who are these digital Indians and how will digital technologies change their lifestyles? Here's what the Boston Consulting Group (BCG)[1] has to say:

> The first 100 million 'digital Indians' were largely men, urban, educated, earning higher incomes, and typically, young. The 400 to 500 millionth 'digital Indians' are going to be the opposite— rural, mid-income, older, with more women included. This

[1] Abheek Singhi and Nimisha Jain, 'The Rise of India's Neo Middle Class', *HT Mint*, 4 October 2016. Available at: http://www.livemint.com/Politics/HY9TzjQzljCZNRHb 2ejC2H/The-rise-of-Indias-neo-middle-class.html (accessed on 17 July 2018).

digital democratization will have a profound impact on how Indians see, select, study, spend, save, socialize and sell.

Despite all the current evidence and predictions, there still remains this doubt about whether digital has made or will make any difference to the teeming lower-income masses and vast groups of people on the margins of society in India. Push that question into the consumption area, and the question is whether digital adoption has made or will make any difference to those that buy and sell at the Jama Masjid, Chandni Chowk, Chickpet, Burma Bazaar, FC Road, Bapu Bazaar, Serenity Beach Bazaar, Jew Town and other such bustling marketplaces that are strewn across the vast country that is India.

At Jama Masjid, as I walk through the narrow lanes, as always I am struck by the sheer business energy that I witness. Everything of every sort seems to be for sale. Everyone seems to be buying something or the other. Much bargaining happens there. Sellers on their part try and ward off any such negotiations by putting up boards that say, 'Fixed Prices'. The buyers will have none of it, and they continue to try and drive prices down. Although I've not been a frequent visitor, over the last decade I have made my once-in-a-while trips to such markets across various cities. I generally look for leather and metal products in these markets. Thankfully, I have been able to find and pick what I believe have been bargain pieces. My physical marketplace trips have gotten less frequent, especially since the last three years. I have migrated online and now do my hunting for bargain pieces in the digital marketplace. What I have been able to find has been a revelation! It is possible that I am more the exception than the rule when it comes to buying almost anything online. That belief is strengthened as I try and walk as many gullies that criss-cross Jama Masjid. My colleagues seem delighted at what they are able to find. When they spot a sought-after product, they stop to bargain. Plus they do something that opens up a window for me in my digital quest. They click pictures of the products they have zeroed in on, and use WhatsApp to relay it back to their folks in Bengaluru. Soon they are on a video call discussing the merits of the product and the price being charged, and then they turn their phone cameras on to the merchandised products so they can send live video back to their folks. The folks back home update my colleagues on probable prices in Bengaluru. They also do a quick check on various e-commerce sites to see if there are similar products available and prices being charged. One of the products they do this for is a 'silk scarf'. Armed with such digital information, they restart the bargaining. They even pass their mobile phones on to the scarf seller to show him pictures and prices elsewhere, and online. The bargaining reaches a crescendo; the seller seems to be on the back foot. A deal is reached, and the price agreed on is a tad but lower than before. My colleagues seem delighted at what they have achieved.

So am I.

Digital matters! Even in such marketplaces. My further observations at Jama Masjid and studies elsewhere reinforce the fact that many more buyers in such marketplaces are using digital information to strike a hard bargain. My research work over the last few years have shown how buyers are transitioning to various stages of digital adoption that, in turn, is aiding them in their consumption decisions. The majority seem to be combining physical with digital in buying journeys. A minority on either end of the curve has gone almost completely digital or is at the starting stages of testing the value of digital in their lives. Of course, this transition is not just being done in physical marketplaces; online too, the breadth and depth of information available to a buyer is radically altering the way buying and selling are playing out. Truth is that digital apps are relevant to all of marketing and business functions. Digital, without a doubt, has transformed the world of production and consumption. It has altered the marketplace to such an extent that the traditional constructs and models of buyer behaviour are increasingly losing their relevance in deciphering the digital buyer.

My work with buyers and businesses who have embraced digital completely and those who have transitioned to a hybrid physical–digital state has revealed shifts in what I call 'power equations'. Although I have spent more time studying the consumer, the spread of digital technology, about which I will elaborate later, has impacted business and marketers in myriad manners.

POWER SPREADS

The monopoly over information or knowledge-driven power is over—in the marketplace and elsewhere, which means that the ability to generate, access and use information is now available to almost anyone, anywhere. The spread of such knowledge is what is removing power, which was traditionally wielded by a few, from their hands and moving it to many others who have never had such access in the past. In the consumption arena, such a spread of power of knowledge to various stakeholders is due to a digital marketplace. To comprehend fully what a digital marketplace is and how it has transformed even the basic acts of selling and buying, you have to ask someone like Shuvankar Maitra. Shuvankar is a freelance painter based in Kolkata. I first chanced on his works while looking to buy art as decor for our home in Bengaluru. I was simply blown away by his works that were featured at online art retailer Mojarto, run by the TV news company, NDTV. Mojarto claims to be the 'the largest online platform for original Indian paintings, prints and collectibles'.

When I was led to Shuvankar's page following a Google search link, I came across four of his brilliant artworks. His work is now open literally 24 × 7 to the world via digital marketplaces. Speaking to *Hindustan Times*,[2] Shuvankar opened up about how virtual

[2] Manoj Sharma, 'Online Gallery: Internet Gives Artists a New Marketplace', *Hindustan Times*, 17 September 2016. Available at: https://www.hindustantimes.com/delhi-news/online-gallery-internet-gives-artists-a-a-new-marketplace/story-UYutypVe Dpg1bnboQDVkxM.html (accessed on 17 July 2018).

display of his artworks has worked for him. This was after his work was patronized by a buyer from the United Kingdom: 'The buyer was a collector from the UK who saw my work online. The biggest advantage of online galleries is that an artist can focus on art; he does not have to run from one gallery to another to show his work'. In March of last year, Shuvankar sold four of his paintings, earning close to US$4,000 in the process. When I spoke to Shuvankar, he agreed that going online had been financially rewarding for artists. It isn't just Shuvankar who has benefited from a digital marketplace. Anni Kumari, after graduating from College of Art, Delhi in 2013, had a torrid time trying to get bricks-and-mortar galleries to display her work. Her being a newbie, they weren't interested in featuring her. The digital marketplace worked better for this young 'emerging artist' (she was nominated in this category for an award by Bestcollegeart in association with Glenfiddich) who sold four of her paintings in the early months of last year.

The digital marketplace by its very nature allows for inclusions that physical marketplaces don't. When I say 'inclusion', it includes access to sellers too. Meaning, a Shuvankar and an Anni may not make it on to the walls of an elite art gallery, but they can adorn any number of 'digital walls' and be available to anyone from any-where for a purchase. Now that alters the market game completely. It throws into a tailspin conventional understanding of access and operability in marketplaces. It's why I term the digital market-place one that's infinite in its scope, reach and existence. No marketplace allows for an optimal engagement between a seller and a buyer as much as the digital marketplace does—optimal pri-marily driven by the access buyers have, and the reach and scale sellers can strive for. The 'optimality' scenario is furthered when digital allows for a connect between a buyer and a seller with both making decisions to engage after having used the power of digital information to arrive at what could possibly be the best pay-offs available to either parties.

THE DIGITAL LANDSCAPE

Until now, businesses have grappled with drawing boundaries for their industries and categories. The fallacy of a limited boundary where competitors are those that make similar products or services has been exposed time and again. The emphasis has been on looking at boundaries in a manner where those others in business that respond to the same buyer needs or problems are included as part of competition. Digital is turning such wisdom on its head by pushing market boundaries even further. Such a widened boundary enabled by global digital access includes almost everyone from everywhere including buyers. The pushing of digital market boundaries to include anybody means that even buyers with 'used goods' to sell can turn into competition. Buyers looking to sell used products actively compete in digital marketplaces with traditional retailers for patronage from a target set of potential buyers. The aiding of customer to customer into mainstream selling has been made possible by the digital marketplace. The fragmented nature of such selling puts it below radar; however, that shouldn't take way from its potential as realized by three young men on the south coast of England in Bognor Regis. Their original act somewhere in the year 2003 was that of selling a pile of Barbara Cartland books they had bought from someone who was putting the unsold set into a rubbish bin. The resale they embarked on got them £270. Egged by global possibilities in the world of reselling books, the young men started the World of Books. By 2016, the company had on rolls 500 employees to manage 70 million volumes a year. Their inventory over time has grown from a measly 1,000 to 2 million in stock. For now, they sell 9 million items to buyers in 90 countries from around the world. In the year 2015, the company made a cool £30 million from sales at a profit of £1.3 million. The company claims to be the world's fourth largest independent seller on Amazon. The Ziffit app the company launched in 2014 rocketed their popularity among people looking to sell their used books. This smart app allows sellers to scan the barcodes on books, CDs,

DVDs and games and avail of instantly global markets prices for the used merchandise for that point in time.

Digital marketplaces are pushing the perimeters of industries way beyond what they have traditionally been in the bricks-and-mortar format. They are also radically altering the nature of business within and across business verticals. Martin Hirt and Paul Willmott from McKinsey, in their article 'Strategic Principles for Competing in the Digital Age', write,[3]

> Digitization often lowers entry barriers, causing long-established boundaries between sectors to tumble. At the same time, the 'plug and play' nature of digital assets causes value chains to disaggregate, creating openings for focused, fast-moving competitors. New market entrants often scale up rapidly at lower cost than legacy players can, and returns may grow rapidly as more customers join the network. Digital capabilities increasingly will determine which companies create or lose value.

The authors point to seven trends that will dominate the digital landscape. They include pressures on prices and margins, emergence of competitors from unexpected places, plug-and-play business models, converging global demand and supply, and continuously evolving business models at a rapid pace.

If a digitized marketplace is rife with disruptive threats to incumbents, it is also a hotbed of opportunities for upstarts willing to leverage its capabilities. The disruptive nature of digital means that all industries are prone to its transformational effects. According to the Russell Reynolds' Digital Pulse 2017 report, senior executives at firms that expect rapid digital disruptions in their industry include financial services, health care and industrial companies (Table 2.1).

---

[3] Martin Hirt and Paul Willmott, 'Strategic Principles for Competing in the Digital Age' *McKinsey Quarterly*, May 2014. Available at: https://www.mckinsey.com/business-functions/strategy-and-corporate-finance/our-insights/strategic-principles-for-competing-in-the-digital-age (accessed on 11 September 2018).

Table 2.1 Senior Executives Anticipating Digital Disruption

Industry	Disruption (to Date, in %)	Disruption (Next 12 Months, in %)
Technology	74	81
Consumer	73	76
Financial Services	65	77
Health Care	56	72
Non-profit	55	62
Industrial	51	63

Source: Russell Reynolds Associates, *Digital Pulse: 2017 Outlook & Perspectives from the Market* (13 March 2017). Available at: http://www.russellreynolds.com/en/Insights/ thought-leadership/Documents/R701001-rr-0085%20-%20Digital%20Pulse%20 final%205.17.pdf (accessed on 17 July 2018).

DIGITAL DISRUPTION

According to a 2017 study from the Global Center for Digital Business Transformation, an IMD and Cisco initiative, titled 'Life in the Digital Vortex: The Sate of Digital Disruption', reveals that 'digital disruption is no longer an intangible event that may occur at some undefined point in the future'. The study finds that the five most vulnerable industries to such a disruption are media and entertainment, technology products and services, retail, financial services and telecommunication. The most digitally disrupted industry is media and entertainment with the disruptive impact spanning categories such as print media, social media, TV, music and movies. When it comes to competition, there is now greater intensity of rivalry among industry incumbents that include video content producers, music and print publishers, and content distributors, among others. What about the threat from entry of new players in this industry? That too has grown rapidly with the emergence of new entrants such as Amazon and Facebook (FB), along with an emerging player like Netflix. What has added to the disruptive change across industries is its picking up pace via quicker digital technology innovation cycles, an explosion of well-funded start-ups, and the striding of Chinese giants such as Alibaba and Tencent on to the world stage.

The original digital disruption story quoted most often is that of Wikipedia. Truth be told, it wasn't Wikipedia that first brought Britannica's encyclopedias down. It was the humble CD-ROM encyclopedias from Microsoft's Encarta that first weakened Britannica. Again, Encarta got into people's homes as an encyclopedia only because buyers pitched for home PCs that came with Encarta. In fact, Encarta as an inexpensive multimedia and as a not-so-comprehensive encyclopedia is what helped Microsoft in the selling of Windows PCs to families. It was primarily Encarta that made Britannica go bankrupt by 1996. It's worth noting that Microsoft had approached Britannica for developing a CD-ROM encyclopedia that the latter turned down. The humble CD-ROM went out of reckoning once Wikipedia came on to the scene. As of now, the digital encyclopedia the English Wikipedia alone carries 5,452,147 articles of any length. If combined, Wikipedias for all other languages carry more than 40 million articles in 293 languages of 27 billion words. Its English version alone has over 2.9 billion words and over 60 times as many as the next largest English-language encyclopedia— Encyclopedia Britannica. What are the other notable achievements of Wikipedia? It's the sixth most visited site in the Internet and is available in more than 250 languages. Students at universities today access Wikipedia more than they go to libraries. In fact, library use has declined by 11 per cent annually, and the use of books has dropped 12 per cent. More than one in two students have been found to stop their research if the information they find on Wikipedia is sparse.

Another product that illustrates well how digital has changed almost everything seemingly overnight is watches. Traditional timepieces over time have and are giving way to digital watches. The Swiss watch industry has over the last few years seen a steady decline in fortunes, much of which can be attributed to a shift in buyer preferences for smartwatches. Pebble Technology Corporation was the first one off the blocks to unveil pixelated timepieces. Pebble smartwatches started selling off Best Buy stores from July of 2013.

In late 2016, the company ceased operations and acquired Fitbit, the company that specializes in wearable technology devices. Since Pebble, the smartwatch business has taken off leaps and bounds with the most notable brand with a smartwatch launch being Apple. In contrast, traditional watchmakers have seen their business decline rapidly. Amidst the decline of analog watches, there have been traditional timepiece brands that have been willing to embrace the opportunity digital was bringing by. Take Tag Heuer Carrera Connected for example. Except for the brand's engineers design-ing the case, everything else that's digital comes from technology partner firms. The Android Wear software in the digital timepiece comes from Google; the touchscreen and electronics come from Intel. Many of the traditional watch brands have also tried to elevate the timepiece from just being one in its digital avatar to it being a wearable device that can perform multiple functions.

In the world of digital, a watch competitor doesn't necessarily have to be another wearable device. It can be a digital device that doubles up on job of displaying time, like a smartphone. Now a similar across-category threat may be hard to predict, and ulti-mately hard to counter as in the case of projectors in the movie business. Samsung's desire to kill the movie projector should be worrying to all those brands that are in the projector business. How does Samsung plan to carry out its threat? By entering the first ever cinema LED screen unveiled recently by the company. The display screen measures at a gigantic 33.8 feet and provides for a 4K resolution. Its peak brightness levels, according to Samsung, are 10 times better than what is offered by standard projector techno-logies. The resultant display is 'distortion free' with neatly an 'infinite contrast ratio'. Samsung has also paired the product with a sound system from JBL by Harman. This 'winning' combination aims to deliver an unparalleled sight-and-sound experience to viewers, the way the entertainment content creators would have wanted.

Digital disruption, a constant in a digital marketplace, afflicts the length and breadth of activities both on the demand and the supply sides. From a supply side perspective, digital is altering value chains from within and across firms that are collaborating to create value for their target set of customers. Digital disruption can turn entire industry value chains obsolete over time. That is because a new chain of firms comes in to create value radically differently from their predecessors. It's also equally possible for digital technologies and platforms to change the way business functions are performed. Take supply chains for example. In a survey by Procurement Leaders of a community of procurement executives, it was found that 8 out of 10 procurement organizations were undergoing a transformation that involved the use of digital resources and capabilities. The development in this function include the utilization of big data and the enhanced use of information related to purchases, news and market trends. There is now even a practice of connecting the physical world via the Internet of things (IoT).

A 2015 study by A. T. Kearney and the business school WHU on European excellence in supply chain management revealed that supply chain managers working at leading European firms believe the most critical digitalization levers to be those of IT integration, more comprehensive use of data and paperless processes.[4] The study found that leading firms across industry verticals were heavily investing in digitizing their business models in general and their supply chain functions in particular. Logistics major DHL was relying increasingly on big data to alleviate risks while having employees use data glasses for the picking process which in turn has resulted in a 25 per cent increase in productivity. DB Schenker, the logistics provider, was investing in a digital mobility lab, while airline companies such as Lufthansa and Emirates were expanding

---

[4] A. T. Kearney and WHU, 'Digital Supply Chains: Increasingly Critical for Competitive Edge' (European A. T. Kearney/WHU Logistics Study 2015). Available at: https://www.atkearney.com/documents/20152/435077/Digital%2BSupply%2BChains.pdf/82bf637e-bfa9-5922-ce03-866b7b17a492 (accessed on 17 July 2018).

on their paperless e-freight offering that came with data cleaning for customers.

Here's another example of the supply and demand side impact of digital in the automotive industry. In an article in *Forbes*, Daniel Newman of Futurum Research and CEO of Broadsuite Media Group writes about the top six digital transformation trends in the automotive industry. These include the use of digital sources in the car-buying process, autonomous driving, connected supply chain and improved manufacturing, predictive maintenance, and data security and protection. Digital technologies and platforms at times even seep into the social scene to change norms and practices that have been followed for long. Take the case of arranged marriages in India. There used to be a time when the whole act of an arranged marriage played out in the physical world with severe limitations. It took time, effort and resources to have the prospective bride and groom meet, followed by their families. Moreover, the number of such meetings had to be limited considering they were both cumbersome to enact and time-consuming. Digital incursions into arranged marriage territory have changed all of that. The first causality of the digitizing of the arranged marriage has been the marriage broker who acted as a middleman between the families of the prospective bride and groom. Virtual marriage marketplaces have ensured that families in India can scour through hundreds of profiles of prospective brides and grooms before settling on who among the lot should be contacted. Indian matchmaking sites such as Shaadi and Indian Matrimony have been inundated with hundreds of profiles of men and women that have been put up by either themselves or the families they belong to. The content available on such sites ensures that there's enough information available beforehand. That means the subsequent contact process becomes easier and the chances of a 'conversion', meaning a tying of the knot, is increased. Online matchmaking in India has now been around for almost a decade. Over time, there have been changes in the way the online matchmaking act plays out in the Indian sociocultural scene. Today, online dating is on the upswing with dating apps such as TrulyMadly, Tinder, Matchify and Woo, among others. What about the online dating scene? According to a study by digital research consultancy Mindshift Metrics, almost 67 per cent of Indian singles are in the know of couples who are dating online. Of the surveyed couples in the study, 33 per cent had met online. According to Mindshift Metrics, this number expected to jump to 70 per cent by 2040. Some estimates put the daily swipe on Tinder in India at a whopping 14 million. Although the dating apps may not put a substantial dent to the arranged marriage scene, they are in a way paving way for changes to such a conservative practice. It's also fair to say that the arrival of digital into developing countries like India has had profound impact on the lives of the country's citizens.

There is no doubt that the digital marketplace is shaking up things for both sellers and buyers. The latter are increasingly using the power of digital to exact for themselves the best value there is. In doing so, they have abandoned value propositions they have patronized in the past for those that have leveraged digital to create enhanced value. On the seller side, recognizing the potential that digital provides business firms around the world has been a hard bent on making changes in the way they work so they don't miss the digital bus. It's also important to note that digital is sparing no sphere and is seeping into everyday areas of people's lives. It's impacting the way people live, transact and, of course, socialize. This means that the potential opportunities digital marketplaces open up are tremendous. A change of status quo is to be expected with digital taking centre stage. Also, henceforth, there will be no point in time when existing status quos will hold for long. That is, digital disruptions will continuously ensure that old business models are replaced. The emerging digitally enabled models that will come in as replacements will again in time be subject to disruptive changes. The hyper business activity that digital guarantees should thrill those who adapt and be the downfall of those that are unwilling.

CHAPTER 3

Perceptions Matter

Late last year, the Media Insight Project, an initiative of the American Press Institute (API) and the Associated Press-NORC Center for Public Affairs Research, embarked on a project to assess people's trust in news media. One particular study[1] they did was on news content people received via shares on social media. In this experimental study conducted online, they created a simulated FB post that talked about a diabetes-related health issue and relayed it to a sample of 1,489 American adults. Each participant in the survey was shown the post as having emanated from one of eight American public figures who often share information about health. The list of public figures included Oprah Winfrey, Dr Oz and the Surgeon General of the United States.

What the study concluded was noteworthy. In addition, the results have huge implications for business and marketing.

As part of the study, Americans were shown a simulated FB post related to diabetes and health, with a headline 'Don't Let the Scale Fool You: Why You Could Still Be at Risk for Diabetes'. Note that the original article was written by a professor and had first featured on the Associate Press (AP) website. For one half of the participants in the study, the FB share they received showed the source as AP, and for the other half, the source was stated as a fictional DailyNewsReview.com. Also, one half of the participants received the share from a 'trusted' sharer (identified earlier by the participants themselves), and the other half got the share from a randomly assigned sharer.

---

[1] American Press Institute, '"Who Shared It?" How Americans Decide What News to Trust on Social Media', American Press Institute, 20 March 2017. Available at: https://www.americanpressinstitute.org/publications/reports/survey-research/trust-social-media/ (accessed on 17 July 2018).

The results of the study are telling on how virtual-content consumers in the digital age form perceptions. To netizens, who shared the news mattered more than the source of the content. On digital media platforms, when people receive news from a trusted source, they tend to believe in the facts and the points of view presented, and in its reporting. In contrast, never mind the legitimacy of the news source, people tend to distrust content that comes via a sharer they are skeptical of (see Table 3.1). Now imagine this in the context of a brand. I mean, imagine marketing messages propagated by the marketer versus those shared by neutral sources including existing buyers. It is common knowledge that digital platforms allow for anyone and everyone to become content creators and propagators. In his book *Here Comes Everybody*, Clay Shirky talked about how people, if given the right tools, can get together to perform a task and pursue ends without traditional organizational structures. Now that is true in the digital marketplace context too. In the business of brands, everyone can have an opinion about products and services on offer, and buyers can get together to pursue mutually shared goals. When it comes to a toss-up between the claims of a marketer and the experiences shared by other buyers, guess whose messaging is likely to hit home with potential buyers? What neutral non-marketer sources state and propagate (read: share) about brands will matter more to those who are bound to see them as 'trustworthy'. Now what should really worry brands is what people do with news they receive from a trusted sharer. They pass it on. It's quite possible that the same happens with information received about brands from trusted sharers. I see a double whammy here. The shared information will both influence the receiver's buying decision-making and then move on via subsequent sharing to other receivers, continuing the influence thread. What makes this scenario truly scary is the speed at which such sharing happens on digital platforms, and the coverage it takes over a relatively short period of time.

Figure 3.1 The Trust Response

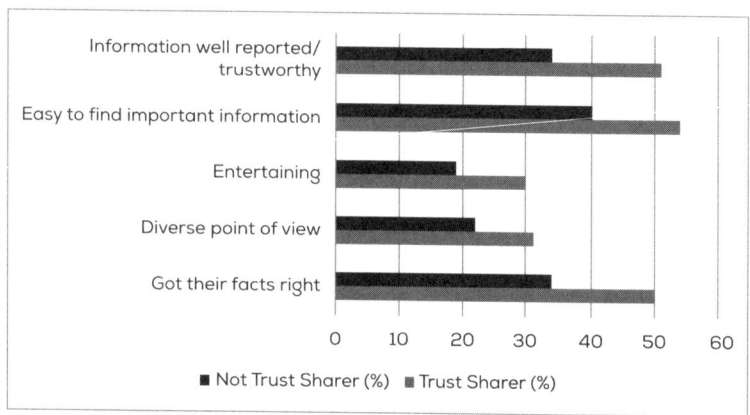

Source: Media Insight Project (Question: How well does each of the following statements describe the article shared in this social media post? Available at: https://www.
americanpressinstitute.org/publications/reports/survey-research/trust-social-media/
[accessed on 18 July 2018]).

DIGITAL INFLUENCE ECOSYSTEM

In the marketing and communication ecosystem, the primary
sources of information and, therefore, influences are marketers
and social reference groups. Who is trusted among the two? There
was a time when the business ecosystem was dominated by marketers' messages. The clutter at such time was caused more by
marketers thronging media platforms with innumerable messages
aimed at their target customers. Not anymore. With digital platforms allowing for buyers and everyone else to voice opinions,
the clutter has been turned on its head. Non-marketer sources
are flooding digital media platforms with content, and therefore
'neutral' clutter. However, buyers today can cut through the clutter
easily and decide on who to listen to, and what content to consume.
In this process of filtration, the entity who is being shut out is
the marketer. On digital media platforms, the people who have the
ears and the eyes of buyers are neutral non-marketers who are seen
as credible, and these include other buyers.

Digital has also changed the way buyers respond to non-marketer influences. Increasingly, buyers are turning to aggregators who bring multiple such influencers on to a single platform so as to aid buyers in their purchase choices. Take, for example, Rotten Tomatoes (RT) and its power to influence movie choices. In 2015, two movies, namely *Terminator: Genisys* and *Fantastic Four*, tanked at the box office. What were their RT score based on aggregated individual reviews by film-goers? The review site gave *Terminator* a score of 25 per cent approval, while at Metacritic (MC), another popular aggregator, the movie performed a wee bit better at 37 per cent. Reviewers on these aggregator sites savaged *Fantastic Four*. Its RT score was a mere 9 per cent. At MC, the film got 27 per cent approval. Both films flopped miserably at the box office. Contrast this with *Jurassic World* and *Furious 7*. At RT, *Jurassic World* scored 71 per cent, and at MC, it performed reasonably well with a 59 per cent score. *Furious 7*, the seventh instalment of *The Fast and The Furious* franchise, scored 79 on RT and 67 on MC. Both movies hit pay dirt of dino proportions. On a budget of US$190 million, *Furious 7* made US$1.516 billion, and *Jurassic World*, on a budget of US$150 million gross, made a whopping US$1.672 billion at the worldwide box office.

It's no wonder then that Paramount's president of worldwide distribution and marketing Megan Colligan said this[2]: 'This was a summer completely designed by reviews and word-of-mouth. I would actually hear people in the grocery store talking about Rotten Tomatoes scores'. What about what established movie critics thought about these movies? PopperlinRosebud, an online commenter, had this to say about movie critic reviews: '90% of movie critics' reviews are utter nonsense. Movie critics write reviews based on what they think will help their own career, which

---

[2] Gregory Wakeman, 'What Studio Executives Think of Rotten Tomatoes', CinemaBlend, 2016. Available at: http://www.cinemablend.com/new/What-Studio-Executives-Think-Rotten-Tomatoes-80747.html (accessed on 18 July 2018).

means praising any liberal movie and unfairly criticizing conservative movies. It's all about being PC'. dion8212, another online commenter, added, 'No one cares about critics. They do not make any difference at the box office, although they go to bed thinking they do'.

The influence of non-marketer sources via digital platforms may not be a new phenomenon. However, what is new is the nuances to such influence and its finer elements. The elements that make up influence in the digital marketplace have not been studied in depth as yet. The gap that exists in the understanding of digital influences means that marketers and businesses are clawing in the dark while making decisions. It's no wonder then that many out there are still persisting with techniques of the past in persuading consumers.

The Media Insight Project study comprehensively proves that even though people claim they care about the source of news, in fact, what matters most to them is the perceived trustworthiness of the sharer. The study results showed that half of the participants who received the FB share about diabetes and health could remember who shared the news, but only 2 in 10 could remember the source of the article content. Again, how significant is that? Imagine that in marketing circumstances. What if consumers can't remember either marketers or claims made by them but can recall opinions shared about brands by 'trusted' people in their social network? Again, what if one such opinion was a negative one about the brand in question?

THE SHIFT IN POWER TO INFLUENCE

The Media Insight Project study findings should make business and marketing sit up and take good notice. If people don't care much about who produces a news article, and instead respond more to who's sharing it, it clearly means that the power to engage has shifted away from the content creator to the propagator. Take a moment and think about that. Content doesn't matter as much as who's relaying it. Now, business and marketing rely heavily on communication. In fact, a major chunk of marketing expenditure goes to communication. Much of that communication budget gets ploughed into advertising and brand building. Companies buy ad and other media space to propagate content to target buyers. They do it over and over again, trying to reinforce their brand messages.

The only problem is that nobody's listening.

Once upon a time, companies thought that the problem could be with the message. Not anymore. It's the bearer who's the bigger problem. With mobile and other technology devices readily available to people, content exposure is essayed in selective manners by the receiver. Meaning, people are curating the content and ensuring they receive only what they want to. That partly means that traditional marketing content is being blocked out. Content allowed in by receivers is required to have a stamp of credibility. The source of such credibility in the digital age is the sender. Who are the credible propagators of content? Surely not marketers! Now I am not saying that there should be no communication that must emanate from the marketer. I am only saying that in the age of digital, it's now time for business and marketing to beware. The buyer is increasingly shutting out marketer messages and is instead opting to engage and respond to content propagated by a social public who are seen as 'neutral experts' and therefore far more credible.

Here's an anecdotal real-life illustration. Thalia lives on her own, away from her family in the city of Mumbai. Originally from Goa, she moved to Mumbai to work as a retail sourcing manager for a reputed lifestyle product retailer. Living alone hasn't been easy, especially when it comes to daily chores such as cooking, cleaning and washing. To make it easier on herself, Thalia uses electronic devices and durables that aid in daily chores. So, for example, on the weekends, she cooks enough to last her for at least half a week.

She stores them in a little refrigerator she bought almost a decade ago. She uses an equally old microwave oven to heat food that she takes out of the refrigerator during dinner time. Just recently, her microwave stopped working. Considering she has had it for a decade, Thalia decided that it was time to get a new one. A bit of search and study online convinced her on considering the purchase of a 20-litre microwave cum grill. Who did she reach out to for help in making a decision? Her close friends who lived in the same city. In addition, Thalia ensured that she read reviews about the brands she was considering as possible purchase options. She found evaluations done on the brands in dedicated review sites. Some of these were written by individuals who had their own review blogs. She also read reviews on e-commerce sites that were selling these brands. She checked their star ratings. Within a week, she was ready to make the buy. During the time Thalia was pursuing her purchase research, she had difficulty managing food at home. She was finding it nearly impossible to heat food from the fridge on the stove and in different vessels. The cleaning up also was piling on. Within a week, she had decided.

When I asked Thalia about whether she was influenced by any particular brand, she told me only to the extent of considering them for a purchase. Beyond that, what mattered to her was what others including experts and buyers had to say about the brands she was considering. Did she add any brands to her consideration set based on information she found on digital media? Yes. Did she eliminate brands from consideration based on reviews and opinions on review sites? Yes.

Again, Thalia and her purchase journey are not something new considering the pervasive nature of digital content and its ability to persuade buyers. However, as I had mentioned earlier, what is new is the journey buyers are making from forming perceptions to learning about brands, and finally the attitudes they develop towards those very brands.

ALTERED PERCEPTION TO ATTITUDE JOURNEY

Perception formation is the first step taken in a buyer's journey. Marketers need to get perceptions right so they can progress buyers towards learning about their brands. This happens when marketing stimuli is reinforced via multiple stimuli exposures. If consumers indeed learn, they soon form attitudes, which means they are now at the cusp of a buy. With digital, the route from perceptions to attitude formation has turned topsy-turvy.

Let's start at the beginning with perception formation. Like I mentioned earlier, for perceptions to be formed, consumers have to encounter stimuli. For brand perceptions, that means marketing stimuli must be selected, organized and responded to. Where are still much of the marketing stimuli being circulated? On non-digital media platforms. Are buyers today receptive to marketing content on such platforms? Not anymore. That's simply because their content consumption has moved firmly away from a yester-year medium like the TV to digital media. The data for content consumption is firmly moving away from traditional media to the digital mobile medium. Note that it isn't content consumption that is down, it's that more people are moving away from consuming content on traditional media. A 2016 Nielsen's quarterly *Total Audience Report*[3] revealed that consumers in the United States were upping their usage of mobile phones and tablets in consuming content. In early 2016, Americans spent more time daily using smartphones to consume media content. In fact, such consumption shot up by 60 per cent from 62 minutes in 2015 to a little over one and half hours in 2016. For the 18–34 demographic, it was found that 39 per cent of media consumption happened on digital devices (smartphones, tablets and PCs), while 15 per cent of their

---

[3] Todd Spangler, 'Americans Are Watching Less Traditional TV as Smartphone Media Usage Booms', *Variety*, 27 June 2016. Available at: http://variety.com/2016/digital/news/live-tv-declining-smartphone-boom-nielsen-1201804202/ (accessed on 18 July 2018).

content consumption came via connected TV devices (game consoles, Apple TV, Roku and Google Chromecast). This was in contrast to 29 per cent of viewing happening on live TV and 17 per cent listening they did at the radio. Another survey[4] by DEFY Media, a digital media firm, found that consumers in the age bracket of 13–24 viewed 12.1 hours of video per week on YouTube, social media and other free online sources, and another 8.8 hours weekly on Netflix and other subscription video services. This meant that the young in the United States were consuming two and a half times more content on the virtual medium in comparison to traditional TV. Video sharing site YouTube topped the list of must-have video sources for the young in America. A whopping 67 per cent of the surveyed young agreed that they couldn't live without YouTube. This was followed by Netflix at 5 per cent and social media services that is an aggregate of FB, Instagram, Snapchat, Twitter and Tumblr at 48 per cent. This is overwhelming evidence that video content consumption is rising on digital media platforms. A mere 36 per cent of Gen Z and millennial consumers said that they couldn't live without traditional TV, according to DEFY's fourth annual *Acumen Report* released in 2015.

What about advertising content on digital media? Here's how young consumers in the United States responded. They were ready to accept advertising in online video depending on the format. A little over half of those surveyed said that they didn't mind watching ads to support their favourite digital stars. While 8 in 10 among the young preferred a 15-second pre-roll ad, 1 in 2 accepted a one-minute spot. Almost 9 in 10 of them opined that a 5-second intro featuring a brand sponsor is always or sometimes OK. Almost 9 in 10 of them approved of product placement in a video (such as digital talent actually demonstrating a product or calling out a sponsor).

[4] Todd Spangler, 'Younger Viewers Watch 2.5 Times More Internet Video than TV (Study)', *Variety*, 29 March 2016. Available at: http://variety.com/2016/digital/news/millennial-gen-z-youtube-netflix-video-social-tv-study-1201740829/ (accessed on 18 July 2018).

Figure 3.2 UK TV-viewing Data by Percentage

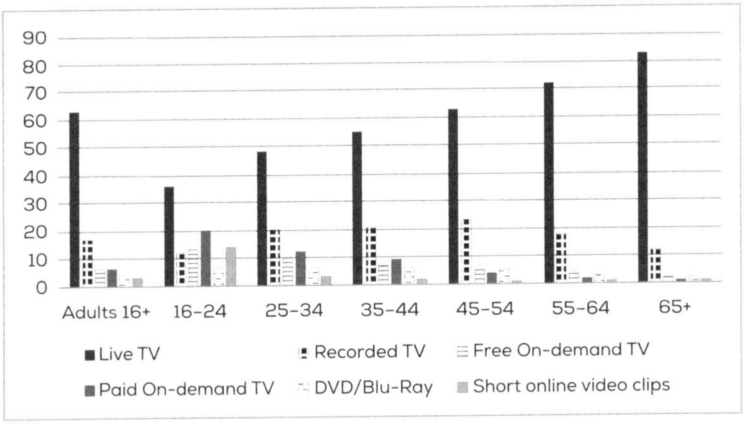

Source: *Ofcom Annual Research Report*, 2016. Available at: http://rls.net.in/wp-content/uploads/2018/02/Trendsetting-Millenials_RAI-Deloitte.pdf (accessed on 09 September 2018).

The shift from traditional media to digital in the consumption of content isn't restricted to the United States. This is now a global phenomenon, especially among young consumers. An Ofcom report[5] about UK TV-viewing habits revealed this (Figure 3.2). People in England watched an average of 3 hours and 36 minutes of broadcast TV in 2015. This is a notable reduction from 4 hours and 2 minutes in 2010. The data meant that there was a drop of 11 per cent in time spent watching TV. Since 2010, viewing on traditional TV has gone down by over a quarter among 16- to 24-year-olds and children. The drop is by 19 per cent among 25- to 34-year-olds. What should worry traditional media broadcasters is the data that revealed that viewers between the ages of 35 and 44 had also reduced their time spent watching TV by a notable 17 per cent in the last 5 years.

The data is clear. TV and other traditional media are being replaced by digital media and devices. This means that people have made a

---

[5] Robert Elder, 'More Young People Are Watching Less Traditional TV', Business Insider, 12 July 2016. Available at: http://www.businessinsider.com/more-young-people-are-watching-less-traditional-tv-2016-7?IR=T (accessed on 18 July 2018).

shift from being passive content consumers to actively deciding on what they want to watch and how much. People in control of their content consumption have major implications of how perceptions form, and beyond. Right from stimuli selection to forming perceptions, everything has changed due to digital media and content. It's now harder for brands to get buyers to passively learn, as the latter are in control of the stimuli they expose themselves to. Add to this the fact that buyers are increasingly being receptive to non-marketer stimuli, and the problem compounds for marketers. For a moment, marketer stimuli is able to break through and form perceptions. What about buyer learning? Agreed that in the case of low-involvement purchases, learning may still be a passive act, but a high-involvement medium like digital can always prompt buyers to move into zones of active learning, even for low-involvement categories of purchase. If consumers are going to actively learn, which surely will happen for high-involvement categories, they are bound to turn to digital content—content they see as far more credible as it is created and propagated by non-marketer sources. Such active learning will then dictate brand attitudes.

Now I am not saying that this is all bad news for business and marketing. I am only saying that business and brands will have to change the way they form perceptions, teach and engineer attitudes in consumers. Let me illustrate this with an example. For years, low-involvement brands have used the associative learning route to get buyers to form perceptions and learn. This sort of teaching worked in an era where messaging was restricted to the one-to-many format. Meaning, the marketer would relay brand messages to a target set of consumers. With the one-to-many format getting replaced by the many-to-many messaging scenario in the digital realm, it's important for marketers to recognize that brand associations can be engineered by anyone and everyone. So a brand like Kellogg's may want to dissociate with Breitbart, an American conservative online news site, so as to preserve their liberal image. The problem is that in the world of digital, there will be a blowback. That's what happened to the Kellogg Company. Using a hashtag #DumpKelloggs, Breitbart went to war.[6] They painted the cereal maker as being anti-conservative. They charged the company with disrespecting and showing contempt to conservative American values. This is what Breitbart President and CEO Larry Solov had to say about the fight:

Kellogg's has shown its contempt for Breitbart's 45 million readers and for the main street American values that they hold dear. Pulling its advertising from Breitbart News is a decidedly cynical and un-American act. The only sensible response is to join together and boycott Kellogg's products in protest.

You see, what Breitbart was doing was associating the Kellogg's brand with un-American values, those of 'economic censorship'. What happened to Breitbart's #DumpKelloggs petition? More than 450,000 signed. As to whether the boycott was directly responsible we may not know, but in the months following, the Kellogg Company has been laying off workers across plants in America. This example is an illustrative of the power non-marketers hold today. They have the power to both rewrite brand associations and influence buyers in making decision choices. All of that means that business and brands have to reformulate the way they position themselves in the digital era. They must also execute on a desired position vastly differently from the way they have done in the past. How this is to be done will be dealt with in the section on digital business and brands.

---

6 Breitbart News, '#DumpKelloggs: Breakfast Brand Blacklists Breitbart, Declares Hate for 45,000,000 Readers', Breitbart, 30 November 2016. Available at: http://www.breitbart.com/big-government/2016/11/30/dumpkelloggs-kelloggs-declares-hate-45-million-americans-blacklisting-breitbart/ (accessed on 18 July 2018).

DIGITAL PERCEPTIONS

Business and brands for long have been using 'interruptions' as a way of exposing marketing stimuli to buyers. When Seth Godin advocated the use of 'permission marketing' over interruptions, he was basing his recommendations as a response to the anger consumers were exhibiting towards interruption content. What Seth was asking marketers to do is to first get permission before supplying buyers with any marketing material. Although such permission taking is still relevant, what I would advocate is to go 'inbound' in getting buyer attention.

The rise of inbound marketing is a direct response to buyers shutting out marketers using personal devices limiting their exposure to content of their choice. That effectively means that perception formation is a goner if buyers decide to shut out marketing stimuli. Here's a pertinent example. When an organic food retailer launched its e-commerce portal in India, it used full-page ads in India's leading newspaper, the *Times of India*. The objective of print media campaign may have been to establish recognition as a natural, environment-friendly retailer among its target buyers. Now the problem with this campaign is one, that is, direct implications on brand perception formation. Note that organic and environment-friendly goods buyers in India are predominantly millennials and they have no time for newspapers. First, here's the data on organic food consumption and growth in India. A recent TechSci Research report[7] on the Indian organic food market predicted a growth of 25 per cent over the next four years. Another report from Yes Bank states that the organic food business in India will increase from a current size of €370 million to €10 billion by the year 2025. Who's doing the bulk of the organic consumption? It's the Indian millennials. Now take a note of their

---

[7] Asian News International, 'India Takes the Organic Path!' Republic, 6 April 2018. Available at: https://www.republicworld.com/lifestyle/fashion/india-takes-the-organic-path (accessed on 18 July 2018).

media consumption habits. Here's what former business head, Digital, BloombergQuint has to say:[8]

> Millennials are definitely into news and rich entertainment and infotainment but their consumption pattern differs greatly from their parents' and grandparents'. They like to get their news and entertainment round-the-clock, through devices they are hooked onto throughout the day, preferably in snackable, bite-sized formats that they can consume while catching a break between meetings and preferably via posts that are peppered with opinions of their friends and family for greater contextual relevance. Millennials are specific about the kind of content they wish to consume and they are specific about how they want to consume it. Unlike baby boomers who often allocate a separate hour to read the newspaper with their morning tea or an evening time slot to watch TV with the whole family, Millennials want their information updates throughout the day. They want to be entertained while sitting at a café waiting for a friend; they want to watch their favourite show from exactly the point where they last left off, whenever they like. They want to know what their favourite celebrity is up to, but not from a famous gossip columnist when they have direct access to the Instagram accounts of everyone from a TV star to their much-loved novelist.

Now guess what that means for the organic food retailer and their full-page newspaper ads? A near-zero exposure is a pertinent conclusion. What does that, in turn, do for perception formation about the organic retailer among its target buyer segment? No impact whatsoever! When newspaper print content isn't even on the radar of millennials, how do you expect them to pick

---

[8] Saket Saurabh, 'Millennials & New Media Platforms: A Match Made in Heaven', NextBigWhat, 27 April 2017. Available at: https://www.nextbigwhat.com/india-millennials-media-platforms-297/ (accessed on 18 July 2018).

such stimuli, organize it, interpret it and move to perception formation?

If millennials are addicted to digital content consumed via their devices, the valid perception formation move would be to grab their eyeballs on such mediums. Now that's where inbound marketing makes a difference. What inbound does through carefully designed and executed content is route the consumer on a digital medium to the material that can arrest attention, form perceptions and capture interest. Furthermore, inbound marketing takes the buyer on a digital engagement journey that can finally culminate in a buying commitment. However, here's a note of caution. It's important for business and brands to track digital adoption by their target buyers across the multiple stages that make up consumption decision-making. An organic retailer shouldn't just conclude that all opportunities to form the required perceptions in minds of millennials are via digital platforms and channels. A 2018 report[9] on Indian millennials and their buying patterns by Deloitte and Retailers Association of India reveals that

> Contrary to the belief of millennials shifting rapidly towards online retailing, they seem to be rather making a gradual shift towards the e-tail mode as brick-and-mortar retailing still remains a relevant channel. Offline channels are specifically preferred for apparel and accessories and personal product categories owing to their emphasis on following major factors, namely, touch and feel of the product, and social connect with friends/family while shopping.

The data from the report showed that the top five reasons for millennials to go offline to shop were touch and feel of the product

---

[9] Deloitte and Retailers Association of India, 'Trend-setting Millennials: Redefining the Consumer Story', Deloitte, February 2018. Available at: http://rls.net.in/wp-content/uploads/2018/02/Trendsetting-Millenials_RAI-Deloitte.pdf (accessed on 9 September 2018).

(76%), to go on outings with family and friends (64%), easier decision-making (63%), the lack of major price advantage online (54%) and to plan product purchases better (52%). The top five reasons why Indian millennials went online to shop were the convenience of buying anywhere and anytime (63%), access to products and brands not available offline (62%), the ability to make frequent purchase (57%), the diverse and wide variety available (56%) and high discounts and lower prices (53%).

It's important for business and marketing to know how to leverage opportunities for digital engagement, keeping in mind the consumption context and exhibited behaviour. In the subsequent chapters, the journey of a digital consumer from perception to learning to attitude formation is explored in greater detail.

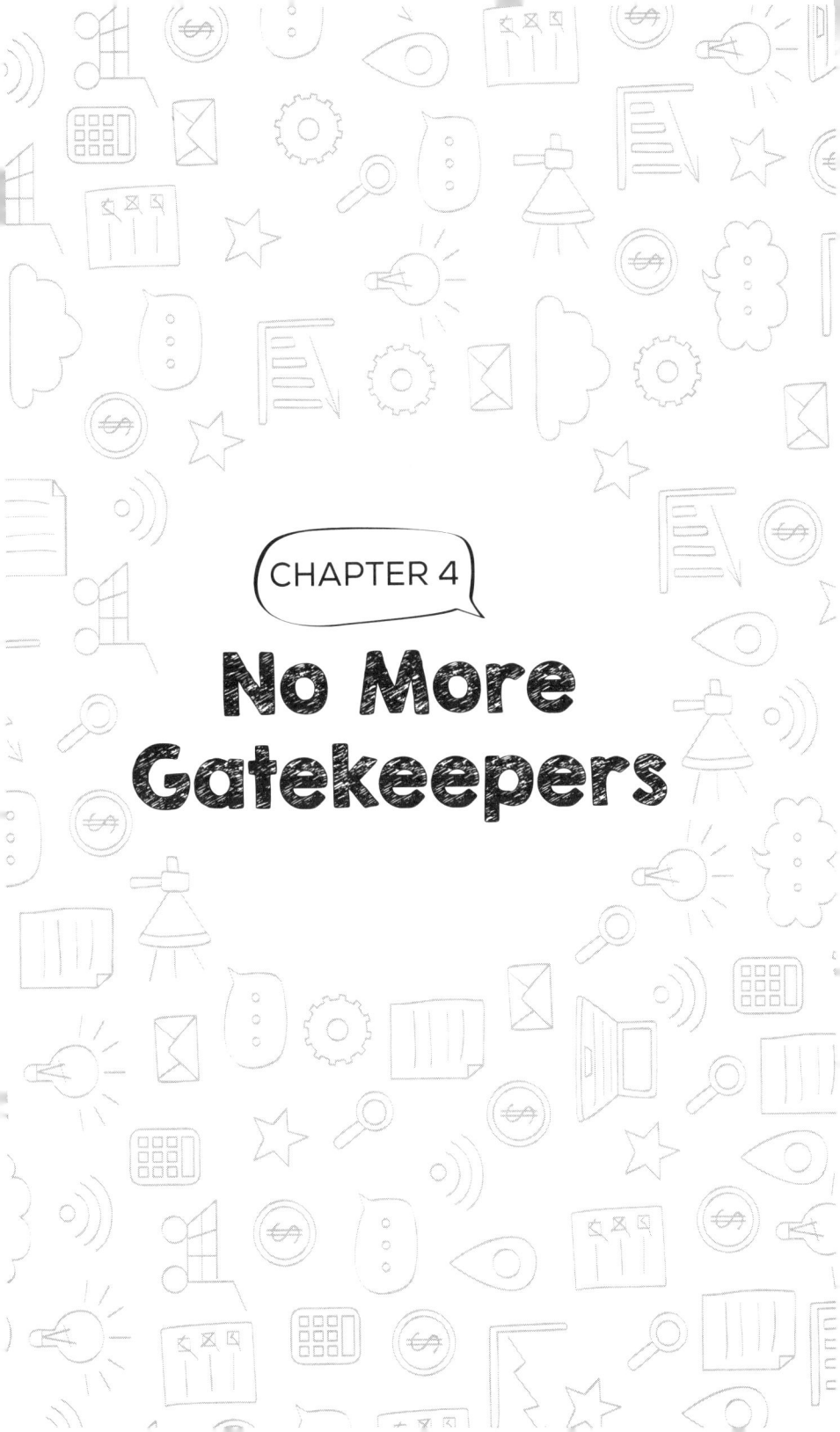

CHAPTER 4

No More Gatekeepers

During the November congressional elections of 2010 in the United States, researchers at the University of California San Diego working with FB embarked on a social influence experiment[1] with eligible voters. On the day of the election, when FB users who were divided into three groups and who had signed up for the experiment logged into their accounts, they were greeted with a message that featured on top of their news feeds. The first group got a message that urged them to vote with a link that led to local polling stations. They were also provided with an 'I Voted' button they could click and a counter that showed how many other FB users had voted. The second participating group too were exposed to the same information when they logged in, with a slight difference. They got to see profile pictures of up to six of their friends who reported that they had voted, plus a tally of how many friends had clicked on the 'I Voted' button. The third of the participating groups got no message. The group that got the message with profile pictures amounted to 60 million. The other two groups numbered at 0.6 million.

The results of the experiment revealed that the group that had seen the message with profile pictures of their friends were 2.08 per cent more likely to click the 'I Voted' button than those who saw the message sans any profile pictures. The researchers ensured that these voters had actually voted by validating claims of one in three of the users by examining public voting records. It was also found that users who had seen the message with pictures were 0.39 per cent more likely to vote in comparison with the other

[1] Rachel Ehrenberg, 'Facebook Peer Pressure Gets Out the Vote', *Science News*, 12 September 2012. Available at: https://www.sciencenews.org/article/facebook-peer-pressure-gets-out-vote (accessed on 19 July 2018).

two groups, whose behaviour had not budged. In actual terms, those that were persuaded to vote by the picture messages numbered at 60,000. Another insight from the experiment was that the 'closeness' of friend links, characterized by a higher frequency and nature of interactions, turned out to be a stronger persuading factor in prompting a vote. Such strong friendship connections users have on FB amounted to 7 per cent.

The FB election experiment reinforced what we already know, that we are most persuaded to act by people we know and trust. However, what was new was the fact that such persuasion need not necessarily require a trusted one to recommend a certain action. A mere knowledge of how people we are close to behaved was enough to have us follow. Here's where digital stepped in. The virtual social platform for the first time makes visible 'others' behaviour'. Meaning, we can now see and witness what others do because of the way we are 'conspicuously connected' on virtual platforms. The 'herd effect' now has greater potency due to enhanced 'virtual visibility'. One of the critical drivers of 'diffusion of innovation' is 'observability'. The innovations in use that we can see have greater chances at adoption by those who see it in action.

THE DIGITAL DICTATE PSYCHE

Technopedia characterizes 'digital influence' as the ability to create an effect, change opinions and behaviours, and drive measurable outcomes online. To understand the potency of persuasion via digital platforms, it's important to have a look at a recast sociocultural environment and its constituent reference groups. From a change perspective, two critical things have happened. One, the nature and propagators of influence in the social sphere have altered. Two, individual psyches subject to these influences are being formed in markedly different fashions from the way it's happened in the past. The traditional forces of persuasion that shape the human psyche dictating the forming of attitudes and beliefs which in turn prompt decision-making have been uprooted by a digitized social environment. That the influencing environment has gone digital in itself is a radical change. Add to that a disruptive change to the nature of the persuader, and you have a potent combination. There are unique characteristics to such an 'altered' virtual environment, such as its borderless and relentless nature that makes its influential effects that much more potent. Yet another feature of digital persuasion is that it is private and personalized. Attribute that to media platforms that allow for personalized access and protection of private spaces, and technological devices that ensure complete secrecy. In effect, this means that individual psyches are being subject to influences that others have no control on. Furthermore, the act of influencing another via digital platforms and using digitized content has near-zero restrictions. Has such back-and-forth digital content with the power to influence scripted a change in the way decisions are being arrived at and made? Absolutely!

Agents of persuasion of the past eras that shaped attitudes and influenced decision-making were limited by geography. Meaning, you could only be influenced by those in your vicinity, and not beyond. A growing-up child would be most influenced by family, community and local culture. That explains why the sensibilities a childharbors would mostly be in line with what was accepted and practised within the community in which he or she grew up. However not anymore. In the digital era, geographical barriers have vanished. That in turn has ensured no barriers to agents of influence from anywhere around the globe. Gatekeeping to keep influence out is also now becoming difficult, and so influencer content is reaching even the very young via digital media infrastructures and devices. A Common Sense Media Research study of 2015[2] titled 'The

---

[2] Common Sense Media, 'The Common Sense Census: Media Use by Tweens and Teens', Common Sense Media, 2015. Available at: https://www.commonsensemedia.org/sites/default/files/uploads/research/census_researchreport.pdf (accessed on 19 July 2018).

Common Sense Census: Media Use by Tweens and Teens' found the use of digital devices and consumption of digital media content to be almost ubiquitous. The study found that

> On any given day, American teenagers (13- to 18-year-olds) averaged about nine hours (8:56) of entertainment media use, excluding time spent at school or for homework. Tweens (8- to 12-year-olds) used an average of about six hours' (5:55) worth of entertainment media daily. This included watching TV, movies, and online videos; playing video, computer, and mobile games; using social media; using the Internet; reading; and listening to music. Tweens averaged more than four and a half hours (4:36) of screen media use a day and teens more than six and a half hours (6:40) a day. A majority of teens (57%) spent more than four hours per day with screen media. (The non-screen portion of young people's media use includes listening to music and reading print.)

Although the quantum of time spent may not be the same, what is common to children around the world is digital content exposure. Such a borderless exposure to digital content is ensuing the elbowing out of traditional localized influence. That children are uniformly turning into digital content consumers and that they are incorporating its influence into their sensibilities is why James Steyer, CEO of Common Sense Media, characterizes them as 'digital natives'. A digitized generation is actively seeking out influences from around the world via digital media platforms and through digital devices, without geographical limitations. So a young student in school, say in India, views YouTube content and plays online games similar to what's being viewed and played by school children everywhere. My son Jaden, who is 10, loves playing the online game Roblox, which is designed and run as a virtual multiplayer social gaming platform. The game which was unveiled in 2006 had grown to a whopping 30 million active monthly players by 2016. Currently, Jaden has around 190 friends who he actively engages with while playing the game. What's more, he has an Instagram account through which he connects to his favourite Roblox friends swapping pictures and exchanging notes. Jaden's growing up is happening as much amidst his family and 'real friends' as with his extended group of 'virtual friends' from around the world. The latter connect and socialize with him via digital platforms. There are a few among his friends whom he idolizes, which explains why he wants to grow up and make it 'big' as a YouTuber!

What such connections shared by children around the world does is that these flatten the world of growing up to make it accessible to kids from everywhere, and thus effectively allow for cross-border influences to seep in and persuade without hindrances. It is true that children still take to friendship groups from within their geographical vicinity, yet it's equally true that they are joining and actively participating in online communities that have no geographical limitations. My daughter Brooklyn, who turned 7 recently, is also a perfect example of a digital native. She takes avidly to online multimedia content that talk of dolls, games, make-up, hair and other such stuff. The YouTubers she keenly follows include Gamer Chad, CookieSwirlC, Lucas and Marcus, Guava Juice and DOLLASTIC. Has Brooklyn's exposure to digital content influenced her choices? It has. Year before last, what she wanted for Christmas was an 'All American Girl Doll'. Last year, she was writing to Santa that she wanted the latest season (incidentally 8) of Shopkins. Now it's important to note that as much as digital content can influence attitudes, individual psyches too can in turn determine the kind of digital activity engaged in. A study examining online habits and personalities, conducted by Soraya Mehdizadeh[3] among 100 FB users aged between 18 and 25 at the York University,

[3] Melissa Hughes, 'Soraya Mehdizadeh, Undergraduate Psychology Student, Finds Facebook Fiends Tend to Be Narcissistic and Insecure', York University, 7 September 2010. Available at: http://research.news.yorku.ca/2010/09/07/soraya-mehdizadeh-york-university-undergraduate-student-finds-facebook-fiends-tend-to-be-narcissistic-and-insecure/ (accessed on 19 July 2018).

found that individuals who were high in the trait narcissism and low on self-esteem tended to spend larger amounts of time on FB. Such low-esteemed narcissists filled their personal pages with more self-promotional content.

In the digital realm, influences that shape psyches, and therefore attitudes and choices, aren't just restricted to the young. A borderless digital world is perfect foil for communities across ages with participants from around the world. A research study by Daiane Scaraboto, Carlos Rossi and Diego Costa on consumer persuasion in online communities found that groups that banded together online turned out to be far more diverse in their make-up. This is an apt illustration of the 'borderless participation' digital platforms facilitate. The study reported that an essential difference between traditional reference groups and online communities was the diversity of participants. Online communities brought together individuals from various social and demographic backgrounds, and those that share interests related to the community topic. In the community examined by the authors' study, participants got together around topics of pregnancy, labour and motherhood. In such online communities, that study revealed that pregnant women found support and advice without any restrictions that came from their place of habitation, and were available 'round the clock' at any given point in time.

Of course, everything isn't hunky-dory about digital influences and our existence in the world of persuasive virtual content. Douglas Rushkoff's warning of 'digiphrenia'[4] is a good place to start at to understand what digital is doing to those of us who are avid participants. He writes,

> As individuals, our efforts to keep up with the latest Tweet or update do not connect us to the present moment, but ensure

---

[4] Jason Cranfordteague, 'Digiphrenia–Excerpt from Douglas Rushkoff's Present Shock', *Wired*, 26 March 2013. Available at: https://www.wired.com/2013/03/digiphrenia-excerpt-from-douglas-rushkoffs-present-shock/ (accessed on 19 July 2018).

that we are remaining focused on what just happened somewhere else. We guide ourselves and our businesses as if steering a car by watching a slide show in the rear-view mirror. This is the disjointed, misapplied effort of Digiphrenia.

The 24×7 connectedness that digital enables ensures that the flow of persuasion is an uninterrupted one. In fact, it's hard for most people to disconnect from the digital realm and exist only in the physical world. Again, at a very real level, digital existences and persuasions that prompt behaviour can have fatal consequences. Take the case of virtual bullying and the disaster it has turned out to be. A study conducted on 4,584 students in grades 3, 4 and 5 by Elizabeth K. Englander, professor at Bridgewater State University in Massachusetts, found that among freshmen at high school, 42 per cent of those surveyed reported that they had been cyberbullied via instant messaging.[5] Among those who reported as having suffered cyberbullying, 37 per cent reported being victimized during high school and 8 per cent reported that they were also or only victimized during their college days. An interesting finding of that study was that of what motivates students to engage in cyberbullying. The surveyed students did not believe that cyberbullies would otherwise engage in bullying behaviour in the real world. This is consistent with observations that suggest that access to electronic communication devices, particularly access to messaging and user-generated web content, draws students into bullying who would otherwise not choose to bully.

INFINITE DIGITAL INFLUENCE

It is amply clear from this that the world of digital and those that populate it are prone to persuade others, and in turn be persuaded. What makes digital persuasion potent and radically different from

---

[5] Elizabeth Englander, 'Cyberbullying & Bullying in Massachusetts: Frequency & Motivations' (Research Brief, Massachusetts Aggression Reduction Center, MARC Publications, 2008). Available at: http://vc.bridgew.edu/cgi/viewcontent.cgi?article=1009&context=marc_pubs (accessed on 19 July 2018).

the way persuasion played out in past is again its 'infinite' nature, in line with what is true at a macro level with digital markets. Anyone from anywhere can persuade anyone from anywhere as long as they are connected via a common digital platform. Also, persuasive content can keep working without any time or geographical restrictions. Moreover, the 'real-time' nature of such persuasion allows for subsequent behaviour to manifest itself in no time, and spread with an unimaginable pace.

Take the case of Sarahah. Created by Saudi programmer Zain al-Abidin Tawfiq in late 2016, the app that allows users to post anonymous feedback entered App Store on 13 June 2017. When the Sarahah website was first launched in the Arab world, its original intention was to allow employees to post constructive feedback to their bosses without revealing their identity. What pushed Sarahah into the 'adoption stratosphere' was when it made its way into the smartphones of millennials and teens in the Arab world. With a zero spend on sales and marketing, the app had trebled its downloads in no time. Within a few months, the app went global. Again, it took the digital world by storm, and within no time hit the number one spot on the App Store across 30 countries. If there was a zero marketing spend that aided the app's adoption and yet if it has had millions of downloads, the persuasive act that prompted users to download and use it must have come from sources other than the marketer. Srinath and Pooja, two among the many young people I talked to, told me that they spotted Sarahah content first on social media. They were at first intrigued, and when they got to know more about the app, they were interested enough to try it out. Soon they were hooked. They also told me about how many of their friends joined in, influenced by their Sarahah posts.

When Seth Godin[6] first talked about spread of ideas at a viral pace in the 'Idea Virus', he was revealing the power of the 'remarkable' to spread at a viral speed in a connected world. What my studies

---

6 https://sethgodin.typepad.com/seths_blog/files/2000Ideavirus.pdf (accessed on 11 September 2018).

on digital persuasion have revealed is that the mundane and the unremarkable can spread too—of course, not at the pace and size of the 'remarkable' but to the extent of its target community. Meaning, 'niche' spreads prompted by persuasive acts within limited audiences too are common to the world of digital. In the case of the latter, what's to focus on is not the quantum of the spread but the enabling environment digital provides to anyone from anywhere to engage in acts of persuasion. It's the 'infinite nature' of digital persuasion that truly makes it remarkable! What, however, truly pits digital persuasion in pioneering territory is its 'democratized' nature. Its ownership is universal. It can never be a monopoly of anyone. It does not discriminate against anyone in that it can be used by anyone without limitations. For the first time, the finances and deep pockets have become irrelevant in the game of digital persuasion. Government and big businesses have at times been powerless against the onslaught of digital persuasion working against them. In India, the current BJP government relied heavily during the previous national elections on digital and social media. The online campaigns the party mounted paid rich dividends in capturing attention, influencing attitudes and finally prompting voter patronage. That was four years ago. Now the party and the BJP government are facing a spirited counter-activism, almost completely digital and run by a handful people—the latter using guerrilla keyboard tactics to hit back against government policies and party ideologies.

What does all this mean for business and marketing?

In the world of digital, as non-business and non-marketer persuasion gain in strength, what should business and brands do? The answer to that starts with a requirement to reconfigure the way business firms design and execute on their persuasive communiqués. Again, this necessitates a mindset shift. Instead of building a communication infrastructure solely dedicated to propagating information about brands and products, businesses instead must redesign and build an enabling digital communication system that allows for users to create and propagate self-generated content regarding their experiences with the firm's brands and other associated assets. Now most companies are unwilling to do this, as they believe that they will lose control over how their brands are presented in a competitive marketplace. Such a worry is a real one. However, that will still not stop users and buyers from using tools, technologies and independent media platforms to propagate self-generated content. So if such propagation is bound to happen, even without a brand enabling it, wouldn't it be better to then join in and support it? Such support will send a message to consumers that the brand is willing to let its users do the talking. If users begin to talk, the rest of the marketplace is bound to listen, especially those who are looking for such information. It's no secret that consumer-generated content is far more credible and, therefore, has greater receptivity. If such content comes off a brand-enabled infrastructure, it gives the business in question immediate access to such content. That in turn can be a treasure trove of learning material businesses can use in real time to get better at the work they do and establish a relationship with the propagator. Remember when such an engagement happens on digital platforms, it's 'transparently visible' to others. The goodwill such acts can generate is tremendous. It shows to buyers thronging a digital marketplace the commitment the brand has towards bettering the value they build and deliver to their customers.

Now all this sounds easy in theory. In the 'real' digital world, executing on brutal transparency is a challenge, both in terms of a mindset and as subsequent action. Business and brands have traditionally been exclusive in that they keep those on the outside engaged until it suits them. Plus any dissonant content is frowned on, and businesses are wired to try and subdue it to make it go away. In a 24 × 7 infinite digital marketplace, never mind all the reputation management, nothing really goes away. In fact, it stays around forever, and so why not reconfigure everything from scratch to build a new way of communication that enables users to be active participants! The building of such an enabling digital infrastructure requires the design to keep user propagation as one of its critical pivots in addition to content generated and propagated by the business or brand. In fact, the design must enable for co-creation of content where the user is as much a participant as the brand owner. So, for example, if it's the information about a launch that's being digitally propagated, the tools, technologies and platforms the brand uses must be open to potential user participation. A real-time co-creation to the extent possible will garner digital eyeballs way more than one (brand) to many (users) relay. Take the Oreo brand for example. At the 2014 SXSW Conference and Festival, Oreo's co-branding worked when it turned its tweets into treats. The millennial festival-goers were able to create personalized cookies based on Twitter topics that were trending and via the use of a 3D printer. Droves of millennials wanting to have a go at this novel concept ensured the ques ran up to almost three hours.[7] Nelson Esseveld at TapInfluence recommends that brands co-create content with influencers.[8] He lists the benefits to brands in doing so, which include authenticity, creativity and multiple voices.

As of now, brands are only beginning to realize that digital persuasion is firmly user territory. It's time they responded to this new reality with a redesigned communication system that is more an enabler than a propagator. Sarahah's widespread global adoption defies traditional segmenting and targeting logic that takes into consideration geographic, demographic and sociocultural variables. Marketers working with outdated persuasion techniques

---

[7] Geoff Gower, 'Marketing to Millennials: The Rise of Content Co-creation', *The Guardian*, 3 November 2013. Available at: https://www.theguardian.com/media-network/2014/nov/03/marketing-millennials-content-creation (accessed on 19 July 2018).

[8] Nelson Esseveld, 'Should You Co-create Your Content with Influencers?' TapInfluence, 22 March 2017. Available at: https://www.tapinfluence.com/co-create-content-influencers/ (accessed on 19 July 2018).

are those with tried and tailor-made persuasive messages for a narrowly qualified target audience. Marketers of now and future are those who know that on the digital platform, acts of persuasion can emanate from anywhere and anyone and persuade anywhere and almost everyone. Again, it's important to remember that digital plays by its own rules. Meaning, persuaders who use digital platforms to their persuasive advantage can either work for a brand or mar its image in an infinite digitized marketplace. A research[9] by Igniyte, a London-based reputation management firm, reveals the dangers of digital persuasion. The study conducted with 1,000 UK-based firms revealed that over half of the lot had been impacted by digital content in a year's time. Three of the four firms surveyed took digital content about them seriously, believing online reviews, comments and forum posts had an impact on their business' financial and reputational status. The same number again believed that malicious content generated about them on review sites may have an influence on their buyers. One in six of the firms surveyed believed that such content had the power to destroy a business. Most of the businesses surveyed did have a clue about what to do about such content and weren't doing anything about possible malicious digital content. Now that's bad news for those that aren't doing anything. That's because such content accessible from anywhere may persuade buyers to stay away from patronizing the firm's products or services.

The case of Sarahah and similar stories of fast-paced digital adoption prove that digital buyers are tuned more into non-marketer persuasive content. Considering that digital media platforms are self-paced by the user, it's only natural that marketer content is first filtered out, or even blocked. By any chance if such content does get through to users, the credibility of such messaging

[9] Igniyte, *The Business of Reviews Infographic* (24 April 2015). Available at: https://www.igniyte.co.uk/reports/the-business-of-reviews-infographic/ (accessed on 19 July 2018).

diminishes in a sea of non-marketer persuasive content that emanates from the user's digital social ecosystem. This effectively means that the marketer has to shift stances in the world of digital. From being content propagators, it makes good sense for marketers to coexist with their buyers in the digital world. Such coexistence, in turn, means that marketers need to listen and respond when appropriate. Moreover, they need to get off their propagator power pedestals and turn into digital beings that are part of a virtual ecosystem. This is not going to be easy. The switch isn't just one of mindset; it's one that requires a complete change of nature. In a recently unveiled research study[10] by the Interpublic Group in association with FCB Cogito Consulting, it was revealed that influencers on digital platforms play a critical role in enabling buyers to make brand purchase decisions. The study spanned across seven countries, namely India, Russia, Brazil, China, South Africa, the United States and the United Kingdom, and surveyed 600 youth, young adults and baby boomers per country. The gender split was even at 50–50. The study showed that both the influencers and those seeking purchase information benefited from an exchange. Those doing the influencing felt that they earned 'respect' in the process and those seeking the information felt more 'confident' about making purchase decisions. Buyers who were supplied with information from social sources felt a greater level of satisfaction with the brands they patronized. The one thing digital buyers looked out for is 'trusted' information. That meant they sought credible information from those they followed as part of their social networks. Those who did the 'credible' influencing did so for pay-offs of being valued and admired. The rise of credible non-marketer influences is why leading brands are making influencer marketing an integral part of their communication and persuasion plans. The data in the favour of the use of influences is

---

[10] Campaign India Team, 'The Rise and Rise of the Influencer', Campaign India, 17 November 2017. Available at: http://www.campaignindia.in/article/the-rise-and-rise-of-the-influencer/441102 (accessed on 19 July 2018).

staggering. RhythmOne's *Influencer Marketing Benchmarks Report*[11] comprehensively proves the value of influencer marketing in dollar terms. The study conducted in the United States found that the average spend by firms on influencer marketing in 2016 stood at US$51,000 per campaign. Those advertisers that executed on influencer campaigns received US$11.69 in earned media value for every US$1 of spend on average. This was an increase of 4.4 per cent of the previous year. The 'engagement rate' across all influencer marketing programmes stood at an impressive 2.01 per cent, a whopping increase of 33 per cent over the previous year's benchmark. Also, the average cost per engagement was tracked at US$0.93 across all influencer programmes. All those advertisers that ran influencer programmes for more than two weeks found that there was an average increase of 14.78 per cent in brand mentions and an increase of 8.73 per cent in positive brand sentiment.

What influencer marketing via digital social sources does is move through the adoption sequence with greater ease and pace. Note that the primary problem traditional marketer face with one-to-many marketer content is its selection by buyers. The latter are conditioned to ignore marketer content, and so to get perceptions to form is a challenge for marketers peddling their own content. However, it is not so with influencer content. The gatekeeping buyers do is loosened when they encounter social influencer content via digital platforms. That means the content gets in without much hindrance, influences formation of perception and, due to the credibility it carries, quickly moves to teaching buyers, thus forming attitudes. That means the journey to attitude formation for digital buyers enabled by influencer content is rapid. That in turn could mean a potential buy.

Marketers hoping to turn on their influences in the digital world are required to harness the power of social influencers in getting

---

their message across. It's when they get off their traditional formats of influence that depend on self-generated and self-propagated marketing content will such a shift to partnering with influencers happen. To truly gauge what non-neutral, non-marketer digital influencers are doing, it's important to step into the world of learning and attitude formation. When buyers learn, they do so either through stimuli exposure or by processing information. Repeated stimuli exposure results in buyers learning passively, whereas when consumers process information, they do so in an 'active', engaged manner. Brands until now, especially those about whom consumers learned passively, have 'gotten away' with engineering attitudes using self-generated marketing content. They have relied on, for example, associative learning through repeated stimuli exposure and reinforcement to teach buyers about their brands. Buyers too on their part have passively learnt from exposure to repetitive marketer content. Influencer marketing has turned that on its head by turning 'passive buyers' into 'active influencers' via digital platforms, and through the use of buyer-generated digital brand content.

One place out of many where such influencing happens is Australia-based Social Soup. The digital influencer community that gets together at Social Soup is supposed to be the largest such digital gathering in Australia. Social Soup promises those who sign up to become influencers a great time as part of a virtual community. This is the promise[12] that Social Soup makes:

> When you join the Social Soup community, you'll be the first to hear about the most exciting new products and experiences. From wine launches and gym trials to the latest shampoo and pizza, all our Soupers (both large and small) are directly involved in shaping the products they care about. Plus, our influencers get to try stuff out before anyone else. #awesome.

---

[12] http://www.socialsoup.com/ (accessed on 19 July 2018).

The 'Soupers' on their part have tested and reviewed products across categories, even those that contain brands that until now have engineered buyer attitudes via passive learning. Take Lipton Iced Green Tea for example. Here's how Tara Leigh, a fellow Souper, reviewed the brand. Giving it 5 stars, she wrote,[13]

> I love it. It is a great alternative to soft drink or sugary juices when you feel like something other than water for a change. It isn't something I will drink every day because of the sugar content (I mainly stick to water) but it is definitely my go to drink when I feel like something different.

What does Tara like in Ice Tea range? 'The refreshing taste! Mango is my favourite'. What would she like to improve? 'Perhaps lower sugar content, although I am unsure how this would affect taste'. Another brand about which Soupers have spoken is Dettol No-touch Hand Wash. Social Soup ensured that 850 mothers were part of Dettol's launch of their No-touch Hand Wash System. These moms got the products in their mail after which they got friends and family over to check out the product. They were also provided with tips for how good hand hygiene could be turned into a fun activity for the kids. Although most Soupers seem to think that the Dettol product is a fun and useful one, Toast, a fellow Souper, throws a spanner in the works. Toast gives the product 2.5 stars and writes,[14]

> The concept is good, but ideally you need to have sensor taps rather than regular ones, otherwise you're still going to have germs all over your taps anyway. I would prefer to use regular hand wash over anti-bacterial ones as the antibacterial ones are not needed and generally cause more harm than good.

[13] http://www.socialsoup.com/review/LiptonIceGreenTea/?title=Lipton+Ice+Green+Tea (accessed on 11 September 2018).

[14] http://www.socialsoup.com/review/dettol_notouch/?title=Dettol (accessed on 11 September 2018).

Toast likes that the product is 'very easy to use and the size of the unit makes it hard to miss and therefore hard to forget'. As an improvement, Toast would like to 'have a feature that prevents multiple uses within a certain period of time, to prevent kids going back for more and more, and wasting the product'.

Numerous others like Tara, Toast and other mums have digitally 'spoken out' on various brands they have tested. They have leveraged their status as 'genuine users' to spread the digital word to those others who are interested. Again, there are numerous such influencer portals out there in the digital space providing for a platform to buyers to spread the 'good' or even the 'bad' word about brands to a digital public. In the process, what they have done is flip buyers from being mere passive recipients of marketing content to becoming active engaged learners heeding to credible non-marketer content. The digital ecosystem on its part has enabled such a flip by allowing for real-time creation and propagation of digital content by buyers to potential users who are clued in on such content of their personal mobile devices. Digital's ability to ensure propagation and receipt of customized, curated influencer content should be seen as a gargantuan opportunity by marketers.

Digital influencer content begs the question that whether all brand learning on digital platforms can be 'active' pursuits by buyers or potential buyers. That question can be better answered by turning it around and asking a counter yet related question. That is, are digital media platforms conducive to passive learning? Media by its nature can be segregated into those that are externally paced and those that allow users to pace themselves and, therefore, are termed self-paced. Users access and engage with digital content at their own pace. They can block and filter out irrelevant content and zero in on digital content that suits their purposes. In a report titled[15]

---

[15] World Economic Forum, *Digital Media and Society: Implications in a Hyperconnected Era* (World Economic Forum, January 2016). Available at: http://www3.weforum.org/docs/WEFUSA_DigitalMediaAndSociety_Report2016.pdf (accessed on 19 July 2018).

Digital Media and Society: Implications in a Hyperconnected Era, prepared by the World Economic Forum in collaboration with Willis Towers Watson, digital media users are characterized as being 'more active'. The report states, 'While traditional media is consumed largely passively, consumers now have enhanced opportunities to share content, engage with content creators, participate in content or even facilitate or sponsor content creation'. UM Wave's social media research study 8,[16] titled 'The Language of Content', which surveyed 50,012 respondents from 65 countries, revealed that virtual users engage with content on digital platforms primarily for purposes of social interaction, expression or recognition, entertainment, information and learning, and work. That the virtual platform allows for a back-and-forth sharing of content means users are emotionally invested in such acts and are therefore highly engaged. That in effect means the learning that happens as a result is an outcome of 'active efforts' put in. The social nature of digital networks means that the 'back and forth' has no limitations whatsoever. That means anyone from anywhere can talk about anything to anyone. That also means that buyer from anywhere can anytime access buying-related information without any limitations. It's possible that buyer may not actively seek out information about certain product categories from digital influencers. However, if that influencer is part of a social network of the buyer, there's a possibility of the buyer receiving 'and' considering information supplied via the network. The smart marketer goes a step further to build such virtual social platforms for buyer communities to form, and then uses influencers to persuade potential buyers to form part of that community. Now this may, at the surface, seem to be a clever play, but it doesn't have to be. This can and must become a genuine effort by the marketer to get to know his buyers and their lives better. In the end, remember that

---

[16] Wave8, 'The Language of Content', UM Wave, 2017. Available at: https://wave.umww.com/assets/pdf/wave_8-the-language-of-content.pdf (accessed on 19 July 2018).

marketing is always about creating and delivering superior value to buyers. How better to create that value than by knowing consumers' lives? Digital is that opportunity to know consumers up close, and to persuade.

For such genuineness to infiltrate marketing and business, it's important that marketers believe that consumers want to learn about brands' products and brands, never mind whether the product category is a low- or high-involvement one. What's of equal importance is that brands shift from engineering brand learning via stimulus exposure and reinforcement. Instead, let the brands genuinely be a part of a social conversation they start and facilitate. Digital allows for brands to engage via acts of information processing. Digital enables conversations, unlike traditional media, which is one-way. Digital media platforms ensure that feedback is instantaneous, which can be worked on without any delay. The back-and-forth, and across-and-away, features of digital mean that social conversations that brands are part of can be designed and executed to include everyone from everywhere. These can be conducted without limitations of time and space. This shift ensures that engaged buyers internalize the 'learning' they undergo. The brand, in turn, sinks deeper into the buyers' long-term memory, allowing for enhanced chances of retrial during purchase consideration.

Imagine that.

The opportunity to teach and learn for business and marketers in the world of digital is humungous. In the next chapter, we will look at how once attitudes are formed, buying decisions are arrived at in the digital realm.

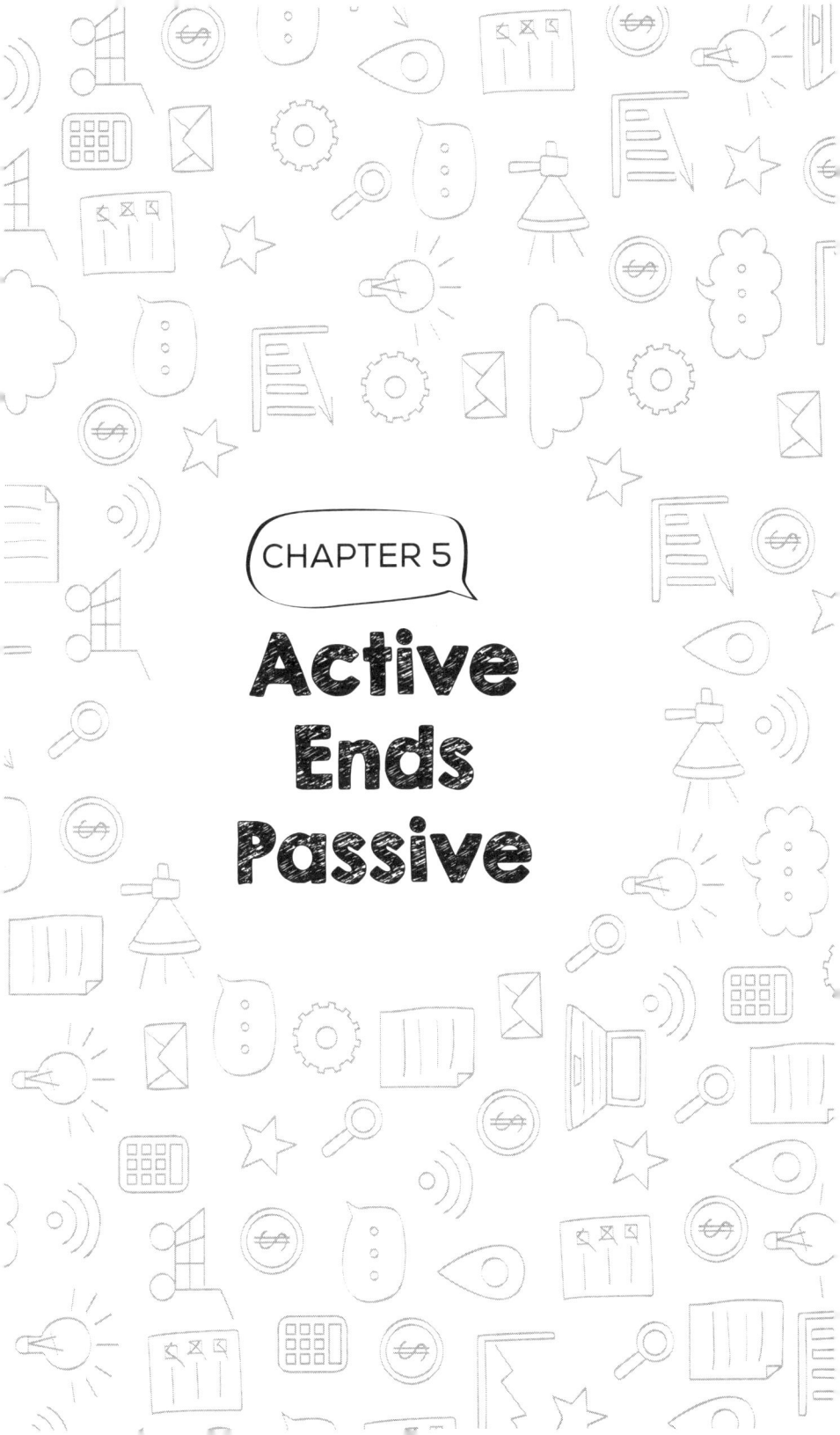

CHAPTER 5

Active
Ends
Passive

It was in the early months of 2011 that the US$19-billion India-based multinational, the Mahindra Group, embarked on an image makeover under a positioning umbrella which was aptly titled 'Rise'. This new mantra adopted enterprise-wide was supposed to be a 'call to action'. This is what Anand Mahindra, the man at the helm of the Group, had to say about the corporate initiative:[1]

> Rise is a simple call to action. We want to encourage people to be part of this movement and to engage with the idea of Rise—to think outside the box, to be inspired and ultimately to take action. A big idea often begins with a tiny spark of inspiration, a spark that can sometimes even ignite a revolution. Through *Spark the Rise* we seek to not only create a platform for 'Sparks' across the country to connect, collaborate and drive positive change but also lend financial support to the best ideas.

The Rise campaign was executed by the New York-based creative hotshop, StrawberryFrog. The campaign communiqués were first unveiled in India and then were to roll out in stages across the globe. The focal goal of 'Rise' was to build a grass-roots cultural movement using digital and offline media. The TV ad that was unveiled as part of the programme was directed by the film-maker Harmony Korine, and the director of photography was the Academy Award-winning Anthony Dod Mantle, the man who had won the award for Best Cinematography 'as a cinematographer' on the movie *Slumdog Millionaire*. All in all, when unveiled, the campaign was impressive in its content and propagation. Over the years since 'Rise' was first unveiled, the Mahindra Group has

[1] Mahindra Rise, 'Spark the Rise with Mahindra', Mahindra, 11 April 2011. Available at: http://www.mahindra.com/news-room/press-release/1313067196 (accessed on 19 July 2018).

embarked on various social and PR initiatives that have borne fruit. The work the company has put in has been truly impactful, to say the least.

In about mid-2017, the Mahindra Group was in the news. The reasons this time were far from desirable. One of the employees at a Group company, Tech Mahindra, had been unceremoniously asked to quit as part of a 'cost rationalization' programme. The employee in question was given no time to leave and was asked by an employee at the human resources (HR) department to resign in a day, or else face a termination. Now this encounter would have made no impact on the outside world, except that this exchange was recorded allegedly by the employee in question. Soon the audio tape found its way on to the Internet and within no time, all hell broke loose. The company came under a barrage of criticism on digital media platforms and social media. The damage was bad enough for the man at the top, Anand Mahindra, to tweet this: 'I want to add my personal apology. Our core value is to preserve the dignity of the individual & we'll ensure this does not happen in future'.[2] The statement of apology was both apt and courageous; however, the story refused to die. In fact, Anand Mahindra's tweet elicited close to 1,800 responses. Some pointed to a new story showing how the CEO, at Tech Mahindra's remuneration, was at a whopping ₹1.507 billion, which was more than the take-home pay of the entire boards of Indian IT majors: TCS, Infosys and Wipro. Another Twitter user posted a story on how the Nintendo CEO cut his pay in half when the Wii U bombed, so his employees would not be penalized.[3] In fact, mainstream media outlets in India picked the story and ran it for a few days. Social media buzzed with the story, and multiple YouTube accounts put up the

---

[2] Anand Mahindra (Twitter post), 7 July 2017. Available at: https://twitter.com/anand mahindra/status/883292235923693568 (accessed on 25 July 2018).

[3] Paul Tassi, 'Wii U Sales Down 36%, Nintendo Executives Taking Pay Cut', *Forbes* (29 January 2014). Available at: https://www.forbes.com/sites/insertcoin/2014/01/29/wii-u-sales-down-36-nintendo-executives-taking-pay-cut/#47772299cf31 (accessed on 25 July 2018).

call audio for fellow netizens to listen to. It almost seemed like everyone now knew about this unfortunate incident.

Now it is possible that there would be no long-term impact of this incident on the Mahindra Group, but it does take a sheen off the whole 'rise' image push. Imagine caring for a grass-roots movement but not caring for your own? Here's the bigger catch. The digital marketplace is 'infinite' in its existence. Meaning, stuff that finds its way into such a virtual market sticks around forever. So in the case of the Mahindra Group, for example, as long as YouTube remains, the HR–employee back-and-forth audio clip will stay around. This in turn means that it can be accessed by 'anyone' at any time. A search on the video-hosting platform for the brand can throw up this not-so-flattering piece that can be detrimental to its image. Now there's a digital brand lesson that can be learned from this episode. To understand it better, it's important to start right from where perceptions are formed by buyers, followed by learning that leads to attitude formation. The way this sequence plays out in digital marketplaces is what is changing the way buyers respond to brands in a virtual marketplace.

THE LEARNING SHIFT

Business and brands always try and ensure that their target audiences are 'positively inclined' to them. Such inclinations are illustrative of an attitude that ensures the brand stays in consideration for a probable purchase as and when buyers embark on a buying decision. Brands use relevant communication tools to build the right images and exact the right attitudes from their target buyers. Now the way this happens is when brands teach their buyers to harbour the right feelings and beliefs about themselves. The Mahindra Group on its part had such an objective (in addition to others) when it embarked on its image makeover with the 'Rise' campaign. There are multiple ways in which brands can do the 'teaching' for their sets of buyers so as to elicit the right attitudes. One of such ways is through the use of 'associations'. The world of associative learning leans heavily on two forms of stimulus–response methods, termed classical and operant conditioning. The most known and valid work in the area of classical conditioning was done by the Russian physiologist Ivan Pavlov. In a study he was conducting on the physiology of digestion in dogs, Pavlov observed that the dogs began to salivate in the presence of the technician who used to feed them, even though the food wasn't around. To experimentally test his observation, Pavlov played out a sound (a stimulus) to the dogs before he fed them. After a few repetitions, he found that the dogs began to salivate at the sound, thus confirming the association the dogs had formed between the sound they heard and the food they got.

Brands direct lifestyle-oriented communiqué materials at their target set of buyers so that the latter can form associations between the brands in question and the outcomes they desire. A movie star uses a cosmetic product on screen so that the cosmetic brand gets associated with beauty. A buy-in into such an association between using cosmetic brands and hoping to turn as gorgeous as the stars endorsing those brands is why buyers in droves swear by such chemical formulations. It's no wonder Charles Revson, the founder of Revlon, remarked, 'In the factory we make cosmetics, in the store we sell hope'. The hope Revson is talking about has much to do with associative learning. As marketplaces have turned virtual and with digital platforms allowing for greater inclusion and participation by buyers, the world of brand associations has been turned on its head. In the virtual world, the way teaching and learning and subsequent attitude formation happen has been reconfigured in digital marketplaces.

Now it's not that associative learning has been pushed out of reckoning; it's that such learning isn't happening via marketer-designed associations; that is, all those traditional tools including advertising are doing nothing to teach consumers about brand associations. Instead, it's other buyers and influencers who are doing the 'teaching'. All thanks to digital and social platforms. So an association of the Mahindra Group with a grass-roots movement was rudely pushed aside when an employee put up a conversation that revealed how he was at the receiving end of a threat of being fired from his job by another employee in the HR department. The infamous conversational audio got replicated in no time, and the virtual space was buzzing with netizens sharing and discussing what they felt about the incident. The media too joined in. Almost all mainstream media and news publications covered the incident. TV channels ran shows discussing the issue of 'terminations' at IT firms in India. The general public that was lapping in all this was in learning mode. That is, they were learning differently from what the brand's communiqués tried to teach them about the company. The fact is that all the possible teaching done by the brand using its marketing communiqués was now being unlearnt. There was new learning that was happening via non-marketer sources, which was diametrically opposite to what the brand intended.

That is bad news for marketers and brand builders. If their carefully crafted messages are being pushed aside by a content-consuming public for content created and propagated by non-marketers and that which does not put the brands in good light, there surely is a problem. Let me illustrate the magnitude of this problem by presenting data on video content generated on the following hosting and sharing platforms: YouTube, Vine and FB. According to a Tubular Labs report, of the top 100 videos on each of the social sharing platforms, brand-created videos made up 17 per cent of the video content on YouTube and less than 1 per cent on FB and Vine (Table 5.1).[4] Now contrast this with user-generated content (UGC). UGC made up 32 per cent of the most watched videos on YouTube. On Vine, the number stood at 17 per cent. On FB, one out of two top videos came from UGC. The rest of the video content is driven by social influencers with 250K+ followers. This is more so on Vine where 81 per cent of the top videos are from this subset.

---

[4] Aaron Dodez, 'User-generated Video Content Is Drowning Out Brands: But There Is Hope', Tubular Insights, 7 May 2015. Available at: http://tubularinsights.com/user-generated-content-brand-videos/ (accessed on 25 July 2018).

Table 5.1 Content Generation on Video-hosting Sites: FB, Vine and YouTube

	FB (%)	YouTube (%)	Vine (%)
UGC	51	32	17
Media Company	25	30	01
Brand	01	17	01
Influencer	23	21	81

Source: Tubular Labs, data based on a period of one month.

The study also found that non-branded video content related to fast-food brands tends to be high, and not always favourable. Take McDonald's for example. The fast-food behemoth tried to push positive spins and generate 'uplifting' content around its 'I'm lovin' it' campaign; however, the video on the same theme that generated the largest number of eyeballs belonged to Media Take Out. This particular video features a twerking lady and a man dropping French fries on her while she does her provocative dance movements. The video has this as a note to it: 'So Uhhh .... We Guess This Is That New HOOD McDonalds Commercial!!! Lol'. The response to this content? 3.4 million views, 7,300 likes, 4,300 comments and 32,000 shares. Media Take Out on its part calls itself the 'The Most Visited Urban Website in the World'. A whopping 5,383,533 people like the site and 5,419,421 people follow it religiously.

How's that for influence?

It's amply clear now that UGC is trumping marketer content. The numbers in a March 2017 consumer study, titled 'Hearing the Voice of the Consumer: UGC and the Commerce Experience', developed by TurnTo Networks and fielded by Ipsos, prove just that (Table 5.2). The study was conducted for a sample of 1,070 US consumers who had made an online purchase for a period of 12 months. The study found that an overwhelming 90 per cent of US consumers revealed that UGC was the most influential element in their purchase decisions. Almost one of every four US consumers

Table 5.2 Influence of UGC on Buying Decisions

	Extremely Influential (%)	Very Influential (%)	Somewhat Influential (%)	A Little Influential (%)	Not at all Influential (%)
UGC	24	29	27	10	10
Search Engines	16	30	28	14	13
Promotional Emails	12	15	30	21	21
Display Ads	10	18	27	21	24
Social	10	16	19	18	37
Mobile Messages	06	13	18	15	49

Source: TurnTo Networks and Ipsos, 'Hearing the Voice of the Consumer: UGC and the Commerce Experience', March 2017. Available at: http://www2.turntonetworks.com/2017consumerstudy (accessed on 11 September 2018).

(24%) said that UGC was extremely influential in their decision to buy or not to buy. The top 'extremely influential' ranking that UGC got beat the next most influential marketing tool, namely search engine results, by a substantial 8 per cent. Overall, search engine results (87%) followed UGC as the second most influential element in buying decisions. Promotional email messages came in third with 79 per cent.

The other key findings from the report show that almost one of four female shoppers (24%) considered UGC to be the most influential marketing tool. Shoppers under 30 revealed a greater influence of UGC on purchasing decisions in comparison to older respondents. A whopping 97 per cent of those who were aged between 18 and 29 reported that UGC had an extreme influence, and a substantial two-thirds of shoppers (63%) felt that it created a more authentic shopping experience. When it came to the confidence in making a buying decision, three-quarters (73%) said that UGC increased their purchasing confidence. Nearly two of three buyers (61%) reported that UGC encouraged them to engage with brands.

There is a marked difference in the way buyers engage with marketer-generated communiqués and UGC. It has been well established that the former isn't seen as credible, as the source of the information being propagated is the marketer. Marketers admit that buyers view their messages as inauthentic, and even if they do consume the content, which could be an infographic, a video or an e-book, it still fails to inspire. It's not so for UGC. There is greater believability in the information generated and propagated by a non-marketer. A study by TapInfluence, billed as the world's first influencer-generated content engine, revealed that three of four consumers turned to social networks for guidance in making purchase decisions. The study also found that influencer marketing content delivered 11× higher ROI than traditional forms of digital marketing. About one of two consumers admitted that they relied on recommendations from influencers while making purchasing decisions.

SHIFT TO ACTIVE LEARNING

To understand the shift in the way buyers today learn about brands in a digital marketplace, it makes sense to go where learning happens the most, that is, a classroom. Note that digital isn't just changing marketplaces but also having a significant impact in academia. Until now, in most classrooms, the lecture-based instructional methods used to promote 'passive learning'. The traditional approach to teaching and learning is one where the teacher engages students with a lesson content in the classroom and applies what is taught by asking students to work on homework assignments outside of school. This model has now been turned on its head with the arrival of 'flipped classrooms' aided by digital technologies and apps. In a flipped classroom, students take a prescribed lesson outside of school aided by digital media and then work on apps in the classrooms with the help of the instructor. Simply put, classroom lectures are consumers' outside class, and homework is now done in class. Why this form of teaching and learning works better is because the old model with its passive learning approach has been replaced by a flipped format that encourages active learning in the classroom.

Now how have students taken to the flipped classroom model? Here are the results from 'Speak Up 2013 National Research Project Findings: A Second Year Review of Flipped Learning'. This study conducted by Project Tomorrow and the Flipped Learning Network used data from more than 403,000 K-12 students, parents, teachers, administrators and community members. The results from the study revealed that three out of four middle- and high-school students agreed that flipped learning would be a good way for them to learn. A significant 32 per cent of those students strongly agreed with the idea of flipped learning. When instructors were asked about the use of digital material in their classrooms, almost one out of two said that they were using videos they found online as part of their teaching in the classroom. Of the lot, 16 per cent said that they were regular in creating videos of their lessons or lectures for their students to watch.

If the digital classrooms of today have 'flipped', so have digital marketplaces. Yesteryear marketer content that pushed passive learning has been supplanted by UGC that engages and ensures buyers learn about brands in an 'active' manner. Now I am not saying that passive learning is irrelevant. Such learning has a lesser chance in driving brand attitudes that may culminate in buying decisions. Instead what is dictating learning is user content available 24 × 7 and that which can be accessed anywhere from any part of the world. This is especially true for high-involvement purchases where buyers actively seek brand information to ensure that they can make the 'best' buying decision. A case in point is Thejas who recently bought a Bluetooth speaker. He started his product and brand information search keeping in the mind his budget of ₹10,000–20,000. He first decided that he needed to know what parameters he must consider while evaluating a Bluetooth speaker. A Google search prompt of 'how to pick

Bluetooth speakers' took him to numerous pages giving him that information. While talking to him, I decided to retrace his digital buying journey. The article he accessed via Google search and the one he read to know more about how to pick a Bluetooth speaker belonged to Eric Mack. The article by Eric was titled 'The Great, Wide Bluetooth Sea: A Bluetooth Speaker Buyer's Guide'. The article contents not only gave Thejas an idea on what to look for while making a buy but also recommended brands he should consider buying if he were looking for specific features in a Bluetooth speaker. To make it easier for Thejas, all the recommended brands were linked to in the article. Thejas read a few more 'expert' articles before moving to an online retailer. His search there threw up a range of brands. Using the filter provided, Thejas narrowed it down to a handful. He then searched for the reviews of these brands first on the retail site and then on YouTube. He wanted to see the brand in use and the rating the YouTube-wide reviewers had provided. When I asked Thejas if he had ever seen a Bluetooth speaker ad on any platform, he answered in the negative. Then I asked, 'Did you visit any of the brands' websites?' 'Yes, a few', he replied.

What brand did Thejas end up buying? A Bose SoundLink Mini process at ₹16,000+. I asked Thejas what got him to settle on the Bose SoundLink Mini. He pointed to a CNET video review of the speaker by David Carnoy, Executive Editor/Reviews at CNET.

LOW TO HIGH INVOLVEMENT

Now it may seem like UGC makes a difference only when it comes to high-involvement purchases, where there is a considerable amount of risk associated with a purchase decision. That's a misplaced notion. Digital media platforms are the perfect foil for those marketers who want to get their buyers actively engaged, even though the product in question may be categorized as a 'low-involvement' buy. Tom Doctoroff in his article titled 'The

False Divide Between Digital vs. Traditional Media' suggests this possibility.[5] He writes,

> Products exist along a low- to high-involvement continuum. Low-involvement products can never become high-involvement ones—a cookie will never be as important as a car—but the goal of all marketing is to nudge the perceptions of a brand from a relatively lower to relatively higher point. This is done with effective positioning, imbuing functional attributes with emotional meaning. When 'shining hair' evolves into 'silky hair that turns heads', or 'soft skin' is elevated to 'soft skin he loves to touch', the wording strengthens the relationship between consumer and brand. Then manufacturers have the luxury of raising prices and therefore profit margins.

Digital is that platform where such a 'nudging' happens and can happen. Smart marketers leverage the inherent power that digital media platforms possess to engage, and they use it effectively to reel in buyers and have them invest time and effort while on a low-involvement purchase journey. The ALS Ice Bucket Challenge that went viral on social media illustrates how a cause can ignite buy-ins if it engages its target citizens in the right manner. In a brief period of two and a half months between June and August of 2014, the campaign prompted 28 million people to join in the challenge. Over that period, they posted content, commented on or liked a challenge post, and put up 2.4 million videos. The hashtags and phrases that were trending during that time included #icebucketchallenge, #alsicebucketchallenge, #strikeoutals, 'Ice Bucket Challenge', 'ALS Ice Bucket Challenge' and 'Strike Out ALS'. What was the outcome of all that 'engaged buzz'? In the year 2015, according to the Chief Communication Officer of the ALS

---

[5] Tom Doctoroff, 'The False Divide Between Digital vs. Traditional Media', *Huffington Post*, 12 May 2015. Available at: http://www.huffingtonpost.com/tom-doctoroff/the-false-divide-between_b_8730234.html (accessed on 25 July 2018).

Association, Carrie Munk, the campaign raised US$5.7 million by the end of July across all chapters.[6] This was almost five times what they had raised the previous year for the same period. The fund contributed came from more than 100,000 new donors, all of whom were roped in, thanks to the Ice Bucket Challenge going viral.

The power of digital is also the power to engage buyers and turn what is traditionally a low-involvement purchase scenario to a high-involvement one. Now this transition has to be approached and understood with care. Note that it isn't products or product categories that are either 'high-involvement' or 'low-involvement' ones. The levels of involvement have to be assessed from a buyer's perspective. Meaning, it's buyers who are either 'highly' or 'lowly' involved with products and categories. The level of involvement is dictated by the level of risks associated with the purchase of a product or service, as perceived by the buyer. So it's possible that the same product has different levels of involvement from different buyers, thus dictating varying levels of learning pursued. To illustrate the transition that I earlier mentioned, let's consider a product like soap. Also, let's assume it that a certain set of consumers buy soap as part of their routine purchases, meaning the involvement levels exhibited are low. Note that a soap brand like Lux from Unilever came into existence 92 years ago as a mass-market toilet soap. The brand shot into prominence with its pioneering association with female celebrities. Positioned as a 'Soap for the Stars', the brand is now in the billion-dollar club. Its patronage around the world comes from buyers who have identified and internalized the brand's promise of beauty and glamour. The brand Lux is an apt example of how mass media content and the technique of associative learning have been used to engineer a connection between using a soap and turning beautiful and glamorous. The truth, of

---

[6] Brian Braiker, 'The "Ice Bucket Challenge": A Case Study in Viral Marketing Gold', Digiday, 14 August 2014. Available at: https://digiday.com/marketing/ice-bucket-challenge-case-study-viral-marketing-success/ (accessed on 25 July 2018).

course, is that soaps can do nothing about a person's looks or life-style. The soap–beauty/glamour association works because people aka buyers aren't interested seeking out the truth. That, in turn, is because the learning that buyers do about soaps comes from con-tinuous and 'passive' responses to advertising stimuli. Such passive (stimulus–response) learning buyers undergo is what seals the association between a brand of soap and beauty and glamour. Note that associative learning works in low-involvement consumption decision-making scenarios. Where consumers aren't keenly involved in making buying decisions, they rarely seek any brand information in active manners.

Passive learning ensures that there is no active seeking of informa-tion about products and brands by consumers. The limited learning buyers do happens at the point of purchase. When consumers buy, they arrive at a final buying decision after having gone through the stages of need recognition, information search and evaluation of alternatives. When most people/buyers who aren't highly involved with a product like soap buy, they do so trusting completely the information that is being supplied by the marketer. Most of that information is passively internalized by the buyer and retrieved when the need for product is recognized and the information search begins. Again, even if a soap buyer actively seeks informa-tion about the products and brands that are being considered, it is limited to what's available at the point of purchase. Based on a combination of associative learning and limited information pro-cessing, buyers make choices from available soap brands. Once they buy and use the product, they rarely bother to pass on either feedback to the manufacturer or their experience with brand to others in their social circle. Remember, that's all because their level of involvement with the product is minimal.

Enter the digital marketplace and you'll find that the way people are making soap-buying decisions is being turned on its head. When I go to the Amazon India page where one of Lux's variant (Lux

Velvet Touch Jasmine & Almond Oil Soap Bar) is selling, I see that 104 people have rated the soap brand. Almost 6 out of 10 reviewers have given the soap 5 stars. There are 87 positive reviews, and 17 critical ones. Among the positive ones, one of the reviewers has descried the soap as follows: 'Good soap. Good smell. Good feel. Good size. Good value'. On the opposite end, a critical review reads, 'Not good for skin, chemicals harm the skin, go use an Ayurveda soap'. I go to another online retailer in India, Snapdeal, and I find that the same soap has 287 ratings. A majority of 165 of them have given the brand 5 stars. One of the positive review reads, 'Amazing to get Lux Velvet Touch Jasmine & Almond Oil Soap Bar 150 g Pack of 3 at ₹90 only. There's much to enjoy, and no scope to complain. Well packed and timely delivery within 2 days'. An unhappy buyer on Snapdeal says, 'Average quality. Not what LUX is known for….' On Purplle, the soap has an average rating of 4.3 stars coming from 18 ratings.

The feedback and ratings given by online buyers at the click-and-mortar stores they buy from span an entire range of fast-moving consumer goods. If the product in question can be rated on functional parameters, online buyers do so. Take for example the case of the 'Spin Mop with Easy Wheels and Bucket for Magic 360 Degree Cleaning'. Here's what online buyers have to say about the product[7]:

The spin mop is useful because by spinning the mop, it can be drained of excess water before mopping floor. However, one has to dip the mop in the dirty water that accumulates in the bucket. This requires water to be replaced with clean water when it looks dirty. The mop attached to end of the stick takes away the back-breaking job of mopping by squatting as all maids do. Good for self-mopping in relatively small flats when the Bai does not turn up.

Here's another[8]:

> It's a good product and I liked the way it spins and removes the water without making your hands dirty. But 2 flaws I would like to highlight although they are NOT a Showstopper: 1. The handle to drag the entire container is very short so be careful that your container is not filled with water completely, otherwise it will spill. 2. The same, you need to use your hands finally when you are done to completely dry out the mop as the spin would not dry it completely but while cleaning it is acceptable.

Now there are two things of particular note here. One, buyers are generating and displaying feedback about products like soaps and mops, and other fast-moving consumer goods that generally fall into the low-involvement category of purchases. Two, these reviews posted by online buyers are aiding others in making up their minds about buying products online. The buyer who goes by the name 'PY' and who posted one of the reviews on Amazon about the spin mop has been ranked number 673 among reviewers by the online retailer. PY has 214 'helpful votes' which he's accumulated via his reviews. The spin mop retailed on Amazon India currently has 4,032 customer reviews. The product has also generated 247 questions from potential buyers that have been answered. One of the questions asked was, 'Does the mop-bucket have partition for clean and dirty water?' The answer from a buyer was as follows[9]:

> No, it does not have a partition for clean and dirty water. I however use a separate bucket of clean water (without

[8] https://www.amazon.in/Gala-wheels-bucket-cleaning-refills/product-reviews/ B00L6TMCFU/ref=cm_cr_getr_d_paging_btm_2?showViewpoints=1&pageNumber= 2 (accessed on 11 September 2018).

[9] https://www.amazon.in/ask/questions/asin/B00L6TMCFU/ref=ask_rp_reva_ql_hza (accessed on 11 September 2018).

wheels!) with a few drops of floor cleaner in it. Now dip the mop in the fresh water and slosh the floor liberally. Then spin dry the mop on the Gala bucket spinner and then mop up-spin-mop up-spin. Empty the dirty water from the bucket, now you may like to rinse the mop on the rinsing attachment (which is real hard to do compared to the spinning) within the Gala bucket with a little fresh water. Thus you are only mopping up the dirty water from the floor into the Gala bucket and not the other way. This is one way, not the best, but we all do make some compromises don't we? ☺ May you have the patience to explain this to any domestic help you may encounter who is willing to learn and not give up immediately!! Wishing you a very happy new year, and thanks for reading!!

DIGITAL ENGINEERED INVOLVEMENT

It's now been established that online product reviews matter to buyers. In fact, these matter to the extent that these affect a brand's ability to convert an interest into an actual online purchase. In a 2015 study by PowerReviews and Northwestern University's Spiegel Digital & Database Research Center, it was revealed that when it comes to brand ratings, 4.2–4.5 out of a possible maximum of five is the ideal average star rating for purchase probability.[10] Although online buyer ratings matter more for high-involvement category products and brands, when it comes to low-risk, therefore, low-involvement purchases, if the brand is priced higher within its category or is a new introduction into the market, ratings matter. Now this is important to note. If a brand is at a higher price point within a category, digital buyers look to feedback from others who have already bought, to see if the premium being charged is

---

[10] PowerReviews and Northwestern University, 'From Reviews to Revenue—Volume 1: How Star Ratings and Review Content Influence Purchase', PowerReviews, 2016. Available at: https://www.powerreviews.com/wp-content/uploads/2016/04/Northwestern-Vol1.pdf (accessed on 25 July 2018).

justified. Since digital content about products and brands is easy to generate, display and even propagate, buyers spend some time to put their feedback out there based on their purchase knowledge. A 2016 Pew Research study found that 4 out of 10 US adults, at some point, have expressed opinions on social media sites like FB or Twitter.[11] This includes more than half of 18- to 49-year-olds who have given feedback on social media. In parallel, since digital content is easy to access and consume, buyers intending to make a purchase use it to make decisions on which products and brands to buy. The interaction between a potential buyer and someone who has already bought even plays out real-time. That's when e-retailers provide a platform for specific questions to be directed to existing buyers who have a choice to respond. Their responses in many cases may be the single deciding factor on whether a buy goes through or stalls.

That reviews matter isn't new knowledge, but that digital buyers are processing information before making buying decisions even for low-involvement category products is. What should worry business is that where traditionally buyers would see marketing content as 'good enough' and passively buy into it for making low-involvement buying decisions, now they are actively considering information generated by those 'credible' others who populate digital marketplaces. This move from passive to active consideration of information may seem as a threat in that it may be seen as shifting the power equation in favour of non-marketer influencers, but it needn't be so. For marketers to make good of such a situation, they have to first take cognizance of the fact that their marketer-led brand messaging is no longer their ticket to a possible purchase, and that too across product categories. Take brand storytelling for example. The attempt here through various

---

[11] Aaron Smith and Monica Anderson, '2. Online Reviews', Pew Research Center, 19 December 2016. Available at: http://www.pewinternet.org/2016/12/19/online-reviews/ (accessed on 25 July 2018).

brand-centred narratives is to embed the brand deep into the psyche of the buyer. With constant repetition and reinforcement, the brand does its best to stay in the long-term memory of the buyer. The hope is that when there's a purchase consideration, the buyer would retrieve the brand from long-term memory and consider it for a possible purchase. Now it's not that brand narratives generated by the marketers aren't important; it's that in an infinite digital marketplace, consumers have taken to storytelling and building narratives about brands anywhere and everywhere. Their experience-driven narratives travel far and wide to influence other potential buyers.

CHAPTER 6

Bye-Bye Habits

Take a moment to think about this. When Chinese handset-maker Xiaomi put out its Redmi Note 5 and Redmi Note 5 Pro models for sale in India late February this year on Flipkart, a leading online retailer in the country, a whopping 300,000 units were sold within three minutes.[1] The sale had started at 12 noon that day. Two hours later, when the company put its Mi LED TV 4 on sale, the entire stock was sold in less than 10 seconds. The phones that were sold were priced at ₹10,000 and above, depending on the model and specification, and the TV was retailed on Flipkart at ₹39,999. As of now, Xiaomi has become India's leading phone brand with a 31 per cent market share, pushing Samsung into second place.

Like I said, think about that.

There have been reams of material written on Xiaomi and its spectacular comeback after floundering for a while, during which time the company pulled out of countries like Indonesia and Brazil.[2] In fact, Xiaomi's turnaround is almost stuff of legend; however, what seems to have missed people's attention is the frenzy that the brand has caused among its targeted middle-income buyers. The fandom Xiaomi has built up in a short span of time has no parallels. The comparison between the cult-like following a company like Apple has and what Xiaomi commands isn't valid at least from a buyer behaviour perspective. The difference is that Xiaomi targets those seeking value for money products. From a

[1] FE Online, 'Gone in Seconds! Xiaomi Says Sold Over 300,000 Redmi Note 5, Note 5 Pro Units Within 3 Minutes', *Financial Express*, 23 February 2018. Available at: https://www.financialexpress.com/industry/technology/gone-in-seconds-xiaomi-says-sold-over-300000-redmi-note-5-note-5-pro-units-within-3-minutes/1076785/ (accessed on 26 July 2018).

[2] David Kline, 'Behind the Rise and Fall of China's Xiaomi', *Wired*, 22 December 2017. Available at: https://www.wired.com/story/behind-the-fall-and-rise-of-china-xiaomi/ (accessed on 26 July 2018).

Table 6.1 Smartphone Brand Market Share in China

Brand	2016 (%)	Brand	2015 (%)	Brand	2014 (%)
Oppo	16.8	Xiaomi	15.1	Xiaomi	12.5
Huawei	16.4	Huawei	14.6	Samsung	12.1
Vivo	14.8	Apple	13.6	Lenovo	11.2
Apple	9.6	Oppo	8.2	Huawei	9.8
Xiaomi	8.9	Vivo	8.2	Coolpad	9.4

Source: IDC, 2016.

behavioural perspective, what the Xiaomi buyers are doing should confound experts who track consumer behaviour. Note that it isn't just the tech-savvy youngsters in India who are Xiaomi addicts; the older ones have also made a beeline for the brand. The latest being the former CEO and founder of Infosys, N. R. Narayana Murthy; Indian's much-loved businessman is reported to have said[3] this about the Chinese brand in an interview: 'Look at Xiaomi: I don't find it any worse than any other brand, but it is so inexpensive and so much value for money. Why would I waste 10 times the price and buy anything else?'

Although Xiaomi had slipped in China to second place in 2016 (see Table 6.1), in India the smartphone brand is steaming ahead (see Table 6.2). In the first quarter of 2018, Xiaomi shipped 9 million units,[4] which was an impressive 155 per cent jump in annual shipment growth. This has widened the gap between the brand and the earlier number one brand, Samsung. For the same quarter, the Korean brand Samsung shipped just under 7.5 million smartphones, registering a growth of 24 per cent over last year. Overall, the smartphone market grew by 8 per cent to 29.5 million units for the first quarter of the year.

---

[3] Eugene Tang, Chua Kong Ho, and Yingzhi Yang, 'Rise of Xiaomi: The Chinese Start-up Poised to Become World's Biggest IPO of 2018', *South China Morning Post*, 18 April 2018. Available at: http://www.scmp.com/business/companies/article/2142130/rise-xiaomi-inside-humble-chinese-start-poised-become-years (accessed on 26 July 2018).

[4] Danish Khan, 'Xiaomi Widens Smartphone Market Share Gap with Samsung in India: Canalys', *ET Telecom*, 24 April 2018. Available at: https://telecom.economictimes.indiatimes.com/news/xiaomi-widens-smartphone-market-share-gap-with-samsung-in-india-canalys/63891082 (accessed on 26 July 2018).

Table 6.2 Smartphone and Feature Phone Market Share in India (Year 2017)

Smartphone Brand	Market Share (%)	Feature Phone Brand	Market Share (%)
Samsung	24.7	Samsung	20.5
Xiaomi	20.9	Transsion	13.7
Vivo	9.4	Micromax	8.7
Lenovo	7.8	Reliance Industries	8.3
Oppo	7.5	Lava	7.2

Source: IDC, 2017.

Note that in India, it isn't just Xiaomi phones that are selling out in seconds. Another brand in this pack of 'sell-out' phones is Shenzhen-based Chinese smartphone manufacturer OnePlus, which entered Indian in 2014. Its most recent offering in the Indian market was the OnePlus 6 × Marvel Avengers Limited Edition. Unveiled for sale on Amazon India at a price tag of ₹44,999, the phone sold out in a mere 30 seconds. Although the number of units sold wasn't available, the fact that the high-price tag didn't cause a hitch is important to note. Both Xiaomi and OnePlus have picked unique ways to open their phones up for sale. The former prefers the 'flash sale' route, while the latter is known to sell via the 'invite-only' model. There has been a lot of interest in knowing who these phone buyers are. Soham Raninga, editor at *Digit*, thinks[5] that it's the geeks who are doing the queuing up online:

> These phones are mostly being bought by Geeks. These users have knowledge about the latest hardware, are looking for the best specs and are willing to take risks. The common consumer will not bother waiting for a sale to see whether they get to buy the phone or not.

---

[5] Vishal Mathur, 'Why Does a Phone Go Out of Stock Within Seconds of Going on Sale?' Livemint, 6 September 2014. Available at: https://www.livemint.com/Industry/N9u9LgpBPphua8nejlNqCK/Why-does-a-phone-go-out-of-stock-within-seconds-of-going-on.html (accessed on 26 July 2018).

Soham is right about the geeks; however, they make up only a part of the pack that waits eagerly for a flash sale. My studies show the demographic to consist of tech-friendly, young, urban consumers who aren't anywhere close to being geeks. What's more, there are even value-seeking, upper-middle income buyers in 30–45 age range in Indian cities and towns who keenly await the launch and selling of new models in smartphones. Elsewhere around the world too, phones are selling out quickly. The new Silk White OnePlus 6 was sold out across Europe in 24 hours. Again, recently, when the Xiaomi Mi 8 went on sale in China, it was sold out in 1 minute and 37 seconds.

That smartphones are being bought by the dozens is now almost a commonplace phenomenon in India and around the world. However, the reasons that are propelling value-conscious consumers to join up in the frenzy are telling of a new era in consumption decision-making on digital platforms. When consulting firm McKinsey[6] studied the digital decision journey of buyers, it concluded that the traditional funnel metaphor no longer holds. That is, the journey from awareness to familiarity to consideration to purchase to loyalty is no longer the route taken by buyers to arrive at and stay with their purchased brand. Instead, McKinsey concluded that in the digital era, the decision-making process is a circular journey that consists of four phases, namely 'initial consideration; active evaluation, or the process of researching potential purchases; closure, when consumers buy brands; and post-purchase, when consumers experience them'. Here's how the process of brand consideration has changed according to the consulting firm; first, buyers consider an initial set of brands based on perceptions and exposure to recent touch points; second, they then add to the list or subtract from the list brands based on evaluations they make; third, the

---

[6] David Court, Dave Elzinga, Susan Mulder, and Ole Jørgen Vetvik, 'The Consumer Decision Journey', McKinsey, June 2009. Available at: https://www.mckinsey.com/business-functions/marketing-and-sales/our-insights/the-consumer-decision-journey (accessed on 26 July 2018).

buyer picks one brand at the moment of purchase; and finally, after purchase, there is a building of expectation based on experience that in turn is used in the next decision journey.

DIGITAL DECISION PROCESS

My studies on digital buying decision-making tried to elaborate on each of the stages of the 'circular' digital decision journey. What I found can help explain the decision-making behind the buying frenzy that plays out when smartphones like Xiaomi and OnePlus are put up for sale online. Data from a study[7] by QuikrBazaar, an online shopping site in India, reveals that 40 per cent of mobile phone owners want to change their handsets within a year. An equal number is keen on buying refurbished phones, and the trend is especially popular with millennials. The study also found that 75 per cent of those surveyed are willing to spend over ₹10,000 for a smartphone. More than one out of two respondents pointed to the need to upgrade to latest technology and specification as being the prime driver for the move to a new phone. Only 9 per cent reported the change as an outcome of their phones either breaking or being stolen. The fact that makes the study of smartphone buying interesting is that those that are upgrading are doing so to keep up with what's latest in the marketplace. This 'keeping up' has a functional and psychological aspect to it. One part of upgrading, and more so for millennials, is about ensuring they have a phone that has even greater processing speeds, memory, clarity, camera and so on. Upgrading is also about being seen with the latest and coolest handset. My focus groups with millennials confirm what previous studies have found—that millennials are glued to their smartphones and even see the gadget as an integral

---

[7] Writankar Mukherjee, 'Around 40% Indians Want to Change Mobile Phones Within a Year: Study', *The Economic Times*, 9 March 2018. Available at: https://economictimes. indiatimes.com/tech/hardware/around-40-indians-want-to-change-mobile-phones-within-a-year-study/articleshow/63238890.cms (accessed on 26 July 2018).

part of their being. A 2017 study[8] by B2X, a Munich-based customer service solutions company, and academics at the Institute of Marketing and New Media at the University of Munich found that 25 per cent of millennials look at their phones more than 100 times a day versus less than one-tenth of baby boomers, and nearly half of millennials look at their phones more than 50 times a day which is three times the rate of baby boomers. The study also found that

> Global smartphone users won't give up their device for one month, even if they were offered a day with their favorite celebrity (74%), a 10 per cent salary increase (56%), an extra week of vacation (50%), US$1,000 dollars (41%) or a holiday at their dream destination (28%).

This level of attachment to a gadget is an indication of a high level of involvement with the products and the category. Involved buyers are active seekers of information. Such seeking, particularly in this case, needn't be restricted to times of purchase considerations. It can even be at other times and for extended periods. The levels of involvement dictating information search that buyers engage in is in keeping with classical consumer behaviour theory. However, where digital does a turnaround to classical theories is in the way it changes the manners and methods buyers employ in seeking information, the amounts they collect and the time they take. Here's where conventional understanding of buying behaviour takes a tumble. In the previous chapters, I have thrown light on how information is collected and from whom. There's been a radical shift away from responding to marketer-driven information to that which is shared by other buyers and non-marketing sources. Also, information is being actively sought using personal

---

[8] B2X Care Solutions GmbH, 'Smartphone Obsession Grows with 25% of Millennials Spending More Than 5 Hours per Day on the Phone', B2X, 18 May 2017. Available at: https://globenewswire.com/news-release/2017/05/18/987607/0/en/Smartphone-Obsession-Grows-with-25-of-Millennials-Spending-More-Than-5-Hours-Per-Day-on-the-Phone.html (accessed on 26 July 2018).

digital devices. What's also happening is that low and high involvements don't matter with respect to information collection and processing. Traditionally, risks associated and perceived with purchase scenarios dictated involvement. Now, risks matter but have been contained because of the increase in credibility of information. The latter, as mentioned earlier, has happened because information now comes from non-marketer sources. Where risks weren't perceived, say as in the buying of a floor cleaner (low financial risks), now there is a consideration of information that is happening. Such information, for example, about floor cleaners, is available at the digital point of purchase and presented in manners where it both catches the eyes of buyers and prompts consideration. Such availability of feedback from other buyers at the point of purchase is influencing buying decisions. In the arena of the physical store, neither is such information available nor is there time on hand for consideration. In a digital store, because buyers have saved on time that they would have otherwise spent in accessing a physical store and browsing through the aisles, they spend it on information available about brand from previous buyers. After all, it is a matter of clicks and reading from the screen.

DIGITAL DECISION RESETS

In a study[9] by academics at Northeastern University and three other universities that examined the link between social media and shareholder value where three consumer mindset metrics were assessed, namely brand awareness, purchase intent and consumer satisfaction, it was found that online buyers differentiated between social media put out by other consumers and those that were propagated by business and brands. The study found that 'all

[9] Jason Kornwitz, 'When It Comes to Social Media, Consumer Trust Each Other, Not Big Brands', News@Northeastern, 18 September 2017. Available at: https://news.northeastern.edu/2017/09/18/when-it-comes-to-social-media-consumers-trust-each-other-not-big-brands/?_ga=2.129642700.1593677796.1528708376-336971180.1528708376 (accessed on 26 July 2018).

social media posts were not created equal'. When digital content in the form of a company's own tweets, FB posts or YouTube videos were put out, they were likely to enhance brand awareness and customer satisfaction. However, such content had no impact on purchase intent. This was in contrast to content on brands put out by other consumers. Such consumer-generated digital material had a significant impact on all the three consumer mindset metrics, that is, brand awareness, consumer satisfaction and purchase intent. The impact of consumer-generated content must also be seen in the light of reduced purchase risk. The credibility such content enjoys also means that the cognitive effort taken by other buyers can be reduced. Digital aggregators on their part use their retail platforms to connect potential buyers with existing. Such connections can be on pointed issues. For example, within a few months of having bought a laptop bag from Amazon India, I was asked a specific question on whether a 15-inch laptop would fit in such the one I had bought. Amazon had relayed the question to me from a potential buyer. Interestingly, I did own a 15-inch laptop and so was able to answer the question in the affirmative. What such answers from customers can do is aid other potential buyers in purchase decisions. Part of such aid goes towards reducing the risk perceived by buyers when they process information about brands in their consideration set. In the case of the laptop bag, one of the parameters of evaluation was the necessity to fit a 15-inch laptop. My answer must have set that requirement at rest.

What all the credible digital information does is speed up the decision process and push the buyer into the buying zone. What it may also do is eliminate brands in the consideration set from a possible purchase because there's credible digital information that puts the brand in poor light. Such information considered by the buyer runs contrary to what the brand's claims are. In fact, if the brand's claims are pitched against brand experiences of buyers shared on the digital medium, chances are that there will only be one winner. That won't be the brand. The quickening of buying decision-making is also aided by specific information available from non-marketer sources on the evaluation criteria. Take the smartphone Asus Zenfone Max Pro M1 currently being retailed on Flipkart at ₹12,999. The product has until now notched up 15,442 ratings and 3,929 reviews. The scores it has received for four parameters of evaluation on a scale of 1–5, with 5 being the best are as follows: Camera: 3.6; Battery: 4.4; Display: 4.4; and Value for Money: 4.8. Potential buyers of the phone have also asked questions about the product. The most popular among the questions listed is: 'Do the phone SIMs support dual 4G LTE/VoLTE standby at same time?' The answers to the question has been supplied by the retailer and other users. One of the 'Yes' answers has until now received a substantial 2,056 thumbs ups. Recently, I was on the lookout for football shoes for my son, Jaden. I was willing to try the 'not-so-famous' brands such as Nivia and Vector X. I was of course concerned about the quality of shoes from these. When I checked the brand on Flipkart, this is what I found. The Nivia Premier Football Shoes had been rated 4 star (range of 1–5) by 2,453 users. There were 398 reviews about the product. I filtered the reviews based on 'certified buyers'. Here are some of the review I read: 'Classic studds!! But it starts slipping while starting on the grounds. Makes our shots more classic!! Overall good'; 'Shoes are comfortable but print quality is very bad. Print is disappearing. Sole quality is very good a little bit like rubber. I recommend this shoe for beginners but not for pros. The packaging is disastrous without any protection….' The learning for me about the product is comprehensive and quick. I can now make a decision about keeping the brand for consideration or eliminating it.

DEATH OF HABIT

Every brand worth its salt seeks the loyalty of its buyers. Brands also know that all loyalty isn't 'real' loyalty. The spurious sort of loyalty occurs when a buyer repurchases a brand multiple numbers of times, not out of a 'real' commitment but out of inertia. What puts a repurchase in the category of a 'loyal rebuy' is when the buyer has a high level of involvement with the product or brand in question. Inertia-driven repurchases happen because buyers aren't bothered enough to search and process product- and brand-related information to make a decision. They'd rather just go back to the brand purchased the precious time around and save on the decision effort. Despite their differences, both inertia- and commitment-based rebuys are habitual buys. In either case, buyers are reluctant to put in any effort and process information about available products and brand choices. What digital has done is put a spoke in the wheels of habit. By reconfiguring the decision process from a funnel sequence to one where the buyer starts with a limited number of brands and adds and subtracts brands to and from the consideration set, digital has ensured that information gets processed across all decision stages. Where digital steps in to disrupt habitual buying is through the provision of additional brand information other than the one being searched for. This happens in the case of both inertia- and commitment-based buying. Furthermore, with an infinite quantum of information available on digital platforms, buyers are prompted to abandon habits and search for product- and brand-related information that can maximize on the value gained.

The business of digital aggregators is essentially one that prompts the search for information so as to maximize on value for the buyer. Take the case of 'Koken met Aanbiedingen', an app that tracks promotions across supermarkets in the Netherlands. In providing real-time promo information, the app enables shoppers to get the best prices on offer. What's more, the app also offers real-time recipe suggestions on the promotional products that are on offer in supermarkets. In effect, what the app does is persuade people to abandon their fixation on a particular brand or product and seek those that are available as part of supermarket promotions. The incentive here isn't just better prices or combo offers on food products but also a chance at using them as part of home dishes based on provided recipes. Paul Marsden, head of Digital Insight at SYZYGY, sees this as an attempt at value maximization by shoppers[10]: 'Value maximization is arguably the most important and most central concept to understanding what guides consumer behavior (without it, very little consumer behavior makes any sense at all). And digital helps consumers value-maximize by providing more information and more choice'. Back home at Bengaluru, shoppers are making the best use of Amazon Prime Now, which aggregates supermarkets with a shopper's vicinity. In a focus group, when I asked users how they shopped via the Prime Now app, I found that the searches were more product-focused. Although brands too were searched for, they were negligible in number in comparison to product searches. Choices of where to buy from were dictated by the best prices on offer across the aggregated stores. Did the nature of information search change when high-involvement products were in consideration? Yes and no. Yes, brands already patronized were first searched for. However, if buyers came in contact with the information propagated by and about other brands online, the search moved away from being brand-directed to product-directed. Recent contacts with buyers via push notifications, short messages and emails did make a difference when it came to digital search for information. For example, a push notification by Purplle, an online seller of cosmetic products in India, provoked a consideration of a brand different from what the buyer usually patronized. An emailer from a newly launched car brand put the company into the buyer's consideration set when an upgrade was considered.

The convenience and persuasive nature of digital information can prompt a buyer to consider it for buying purposes. What such information effectively does is break into a buyer's habit and force a consideration of multiple brands. If such product- and brand-related information is relayed by a trusted person who forms part of a virtual social circle, the impact it has on a buying decision is acute.

---

[10] https://digitalwellbeing.org/how-to-help-consumers-value-maximise-in-super markets/ (accessed on 11 September 2018).

CONSUMER-ENGINEERED BRAND BUILDING

A half-a-decade-long research study by academics at MIT Sloan School of Management and NYU Shanghai throws new light on how consumer-driven narratives are influencing purchase choices. The field research conducted by the academics centred on two companies, namely BMW (the German automaker) and Suruga Bank based in Numazu, Japan. For both the brands, video content was created featuring consumers telling brand-related stories which were then presented to respondents in test environments that simulated social media sites. Incremental changes to perceptions that participants had about the brands were measured. These included connection between the self and the brand, trust in the brand and consideration of the brand for a purchase. The study outcomes found that there was a 32 per cent increase in brand consideration by participants when consumer storytelling was involved. It was found that brand stories shared on social media that originated from consumers were especially powerful in shaping brand attitudes. Furthermore, those companies that didn't weave consumer stories into their overall brand narrative were missing out on engaging with potential buyers in meaningful manners. The study suggests that the locus of control in the digital era is moving away from a brand on its own to a combination of content created by the brand and the consumer. The arrival of what is termed 'consumer-to-consumer storytelling' is altering the way brands build equity in a digital marketplace. The study found that when consumer storytelling was involved in a brand's narrative, its appeal with the consumer was enhanced significantly to the extent of consideration for purchase.

Multiple research studies over time have established the importance of the source credibility in the persuasive impact of a message. The switch from brand-driven narratives to consumer-driven narratives should been as a shift in credibility from the perspective of the buyer. Add to it the fact that consumer-created content is drowning marketer-created content, and you have a transformational scenario that is influencing the way buyers are making brand buying decisions. What digital product- and brand-related content is disrupting is the propensity of buyers to slip into habitual modes of decision-making. Such disruption effectively means that brand loyalty may be an occurrence of the past in physical marketplaces where buyers neither had the information, or time, nor even tools to make comparisons across product and brand alternatives. In the physical market settings, the effort that buyers had to expend to process information put them off. Not anymore. In the infinite digital marketplace, information processing is a technology-aided endeavour with minimal inconveniences, and the effort required on the part of the buyer is also minimal. It's no wonder then that brand loyalty is in danger. Brands can no longer count on continued patronage pf their buyers. In a study by SKIMS and Mars Petcare that explored 'disrupting the habitual consumer decision journey', it was found that online-dominant consumers tended to switch brands more often than offline-dominant consumers. When it came to a disrupted decision journey, the consumers encountered 1.5–2 times more touch points than habitual journeys with varying touch point patterns and needs. Furthermore, it was found that more than 60 per cent of consumers were engaged in post-purchase activity, following disrupted journeys.

As much as digital is a threat to the incumbent brands who have for long been enjoying the spoils of consumer habit, it is an opportunity for entrants trying to break in and dislodge the loyalty enjoyed by the former. With increased personalization of devices and content, the opportunity to connect one-to-one with buyers is there for the picking as the next chapter reveals.

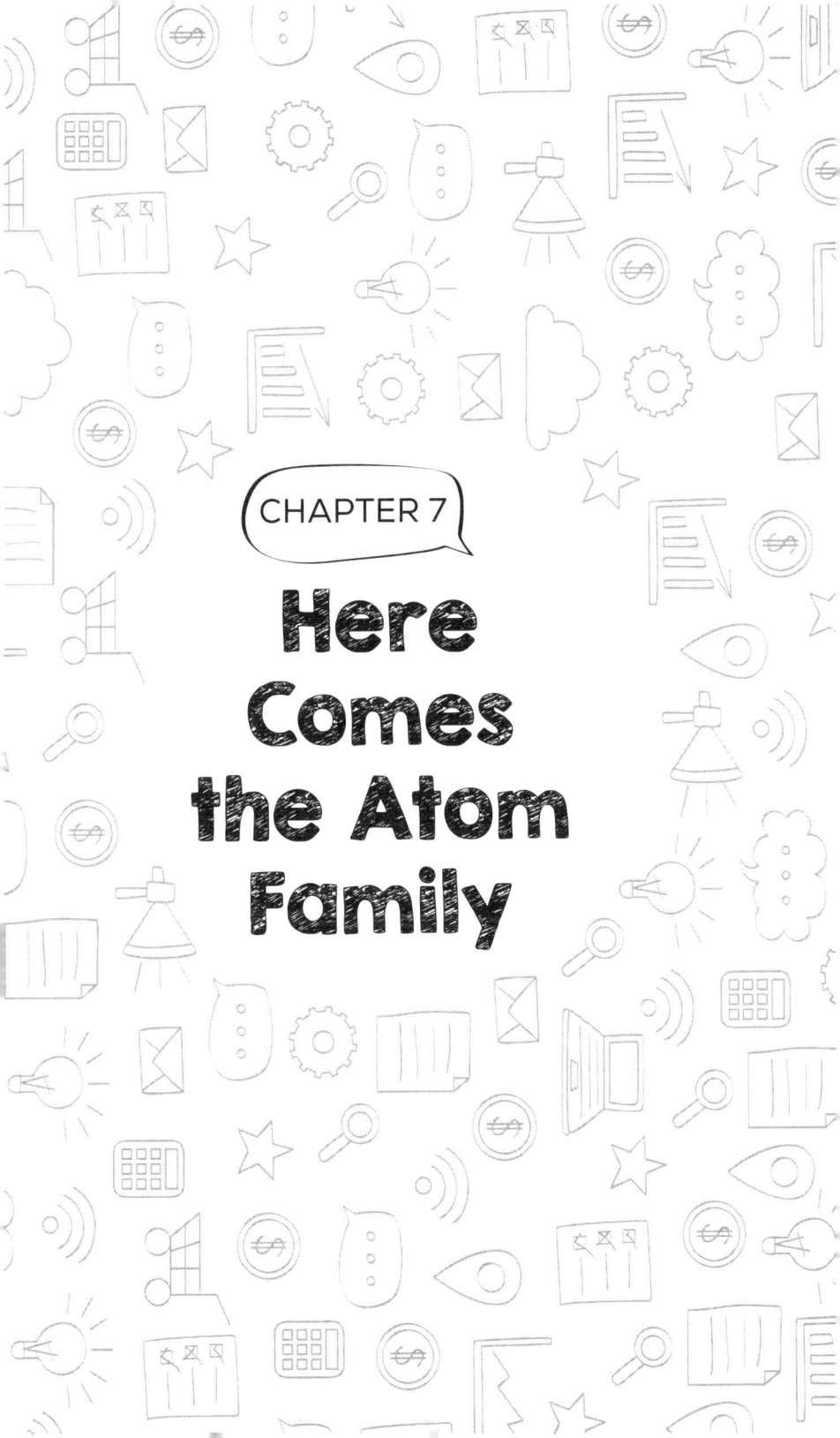

CHAPTER 7

Here Comes the Atom Family

I first came to know of Logan Paul (LP) when my son Jaden decided that what he wanted for Christmas was a 'Maverick hoodie' from LP. Curious, I began my search on who LP was. It turned out that he is among the biggest YouTube stars, and his vlogs on YouTube have raked in millions of views. In fact, he has close to 20 million subscribers who have signed up to his channel on the video-sharing site. Although as parents we do keep a check on the content Jaden follows via his smartphone and tablet, we've never really followed the influences he is being subject to via such content. That he was an LP fan was news to me. When I shared the numbers about the following YouTube stars have at an academic consortium that I attended recently, most people who were part of audience had no clue. They hadn't heard about LP or CookieSwirlC or PewDiePie. The one person in the audience who knew LP had a son who was Jaden's age. Coincidentally, that boy too wanted an LP Maverick hoodie.

To understand digital influences at its place of origin, you have to grasp what is now becoming universally pervasive, namely a digital culture. Here's how Clayton d'Arnault, founding editor at *Digital Culturist*, describes digital culture[1]:

> Digital culture is a blanket concept that describes the idea that technology and the Internet significantly shape the way we interact, behave, think, and communicate as human beings in a societal setting. It is the product of pervasive technology and limitless access to information—a result of disruptive

[1] Clayton d'Arnault, 'What Is Digital Culture?' *Digital Culturist*, 20 October 2015. Available at: https://digitalculturist.com/what-is-digital-culture-5cbe91bfad1b (accessed on 27 July 2018).

technological innovation within our society. It is a lifestyle, and you're part of it. You're living it. Digital culture is the Internet, transhumanism, AI, cyber ethics, security, privacy, and policy. It is hacking, social engineering, and modern psychology. More contextually, digital culture is using social media as our main mode of interaction with others; sharing every moment of your life on the internet; the selfie phenomenon; the live streaming obsession; the anonymity provided by online communities; Apple Pay and Android Pay; wearable technology; the use of emoji to enhance communication; internet/cell phone addiction; the sharing/on-demand economy; cloud computing and storage; the internet of things.

Before the impact of digital culture on the process of socialization is understood, it's important to know why this is happening. A research study by a graduate student on families in Ireland gives us pointers to the same. The study done among families tried to explore the relationship between new media technologies in use within households and social interactions among individuals that made up such homes. The goal of the study by Siobhan McGrath done among households in Kilcock, a town and townland in the north of County Kildare, Ireland, was to see if the use of new media technologies was bringing family members closer to each other, or if it was increasing distances among members, resulting in greater levels of privatization within homes. The three major findings of the study[2] that was titled 'The Impact of New Media Technologies on Social Interaction in the Household' pointed firmly to a spike in the levels of 'social isolation', 'individualization' and 'privatization'. The study found that when the media devices in use were personal in nature, there was greater isolation among members at

---

[2] Siobhan McGrath, 'The Impact of New Media Technologies on Social Interaction in the Household' (National University of Ireland, 19 April 2012). Available at: https://www.maynoothuniversity.ie/sites/default/files/assets/document/SiobhanMcGrath.pdf (accessed on 27 July 2018).

home. The content such devices beamed in were consumed privately and in a highly individualized manner. If I were to take this result and see if it correlates to what happens at my home, I must admit that it happens exactly on the same lines. All of us, when we use our personal devices, are tuning into individualized content in a private manner. It's for this reason that I didn't know a thing about LP until Christmas time.

THE ATOM FAMILY AND DIGITAL SOCIALIZATION

The process of socialization kicks in when individuals acquire and engage with values, norms, beliefs and the general social standards of a particular society and culture. Much of this process starts very early for a child with the primary source of influence being the family. The other sources that participate as influencers include peer networks, institutions such as schools, colleges, and workplaces, community, religion, class and culture. Media is an important entity that is seen as a powerful agent that progresses the process of socialization. The interaction that individuals have with these agents while growing up shape the social norms, values and beliefs adopted.

With the advent of digital media, technologies and devices, the basic nature of the family has altered. In the past, there has been a breakdown of the joint family system, resulting in the proliferation of nuclear families which started in the 1960s and 1970s, especially in urban places. With digital, nuclear families have broken down further to members isolating themselves within. So it's commonplace to see every member of a family huddled up with personal digital devices in a separate place in a modern urban home. I call such an atomization phenomenon as resulting on the emergence of the 'atom family'. In atom families, the process of socialization is playing out way differently from the way it has happened in the past. Specifically, two important aspects of this new way of

Table 7.1 Smartphone Addiction: Hours Spent Online via a Mobile Device

Country	2012	2016
Brazil	1.45	4.48
China	1.4	3.03
United States	1.3	2.37
Italy	1.34	2.34
Spain	1.45	2.11
South Korea	1.3	2.1
Canada	1.22	2.1
United Kingdom	1.29	2.09
Germany	1.25	1.37
France	1.18	1.32

Source: Statista Digital Market Outlook.
Note: Based on Internet users aged 16 years and older.

socializing need to be noted. One, the influence of parents in such settings is weakening, and two, the impact of digital influencers accessed via digital media and devices is on the upswing. The former is happening partly due to a shutting out of the parent by digitally engrossed/addicted children and also because the parents themselves are contributing to the isolation by being 'digitally hooked' (see Table 7.1).

It's a myth that addiction to digital devices and being obsessively glued to digital content are only commonplace among the youth. Data reveal that the older generation is as much, if not more, hooked to their mobile digital devices and content. The adult email addiction illustrates this well. Statistics show that more than one in two employees check their official email during the weekend and before work and after they are done. Adults in the range of 60 per cent log into their emails while vacationing, and a 5 per cent of them check emails while at a funeral. What about while attending a school event? One in 10 adults are guilty of logging into their email at such a time. The practice of using digital devices during meals cuts across all generations (Table 7.2).

Table 7.2 Technology Use During Mealtime

Generation	Technology Use (%)
Gen Z (15- to 20-year-olds)	38
Millennials (21- to 24-year-olds)	40
Gen X (35- to 49-year-olds)	45
Baby Boomers (50- to 65-year-olds)	52
Silent Generation (65+)	42

Source: Nielsen Global Survey.

The Deloitte Mobile Consumer 2016 research study too has some interesting insights on digital dependency. Data from the study revealed that one in three people check their phones within five minutes of waking. That means the exposure to digital content starts even before people get out of their beds. What about the night time? Again, one in three check their phones in the middle of the night, and if it's 18- to 24-year-olds that you're talking about, the nocturnal use rises to one in two (Table 7.3). Here's what seems like a shocker. The study revealed that 8 in 10 people admit to using their smartphones while conversing with friends. A similar number do it while watching TV during evening time.

The rampant use of technology devices across generations is ensuring that members of a family are isolated from one another. Mealtimes, which were opportunities for conversations, are turning into engagements with minimal family talk. All of this

Table 7.3 Night Use of Smartphones by Age

Age Group	Night Use (%)
18–24	50
25–34	48
35–44	37
45–54	27
55–64	20
65+	14

Source: Deloitte Mobile Consumer Survey 2016.

digital-enabled isolation is ensuring that traditional agents of influence that are responsible for the formation and adoption of normative beliefs are being swept aside, replaced by digital influencers. Such new influences even have the power to dictate identity development in adolescents. What sets external digital influencers radically apart from their non-digital predecessors is that they are global in nature. Sarah Jenner and Daniel Suss at the Zurich University of Applied Sciences sum this up well in their article[3] titled 'Socialization as Media Effect':

> With global media and a rapid uptake in Internet penetration, socialization and identity development become more global. Identity construction is a process of integrating both global contents and local cultural tradition (language, religion, ethnicity), and media technology has significantly accelerated intercultural and transnational exchange. Consequently there is a global tendency toward cultural mainstreaming and homogenization. Regardless of the country and its local traditions, people listen to similar popular music (e.g., Lady Gaga, Psy), entertain themselves with similar books, movies, and TV shows (e.g., *Star Wars, Harry Potter, Twilight, Grey's Anatomy*), play similar computer games (e.g., Call of Duty, FIFA), and use similar social apps (e.g., Facebook, Instagram, Snapchat)—all of which are often based on North American culture, which is in turn linked to internationally similar fashion trends and assimilates beauty standards, converging food preferences, and similar celebrities. Conversely, cultural stereotypes tend to get reinforced as part of local identities in the midst of converging cosmopolitan identities.

In atom families, it is now virtually impossible to do any gatekeeping to stem the impact of digital agents participating in the process

---

[3] Sarah Jenner and Daniel Suss, 'Socialization as Media Effect' (John Wiley & Sons, 2017). Available at: http://sarah.genner.cc/uploads/5/1/4/1/51412037/socialization_as_media_effect.pdf (accessed on 27 July 2018).

of socialization. Writing[4] for almost two decades, Marc Prensky had coined the terms 'digital natives' and 'digital immigrants' to throw light on how the arrival and rapid dissemination of digital technology had altered students in US universities. He had pointed that the average graduate students living in the digital age had spent twice the amount of time playing video games than reading books. According to him, college grads in America were making computer games, email, the Internet, cell phones and instant messaging an integral part of their lives. What impact did this have on the students according to Prensky? He claimed,

> It is now clear that as a result of this ubiquitous environment and the sheer volume of their interaction with it, today's students think and process information fundamentally differently from their predecessors. These differences go far further and deeper than most educators suspect or realize.

I agree wholeheartedly with Marc Prensky. Working with college grads for the last one and a half decades, I've seen, observed and studied this phenomenon unfold among Indian students over time. As a father of two, I can see how my children are growing up and imbibing social values and standards markedly in a different fashion from how my brother and I did. Now if digital influences are shaping people's psyches differently, that must then extend to the attitudes they form and harbour, and the behaviour they exhibit. A research study from a decade and a half ago by Christina Lee and Denise Conroy at the University of Auckland and Cecilia Hii at the Thames Business School, conducted on a sample of adolescents aged between 13 and 19 years, and who had an Internet

---

[4] Marc Prensky, 'Digital Natives, Digital Immigrants', in *On the Horizon* (MCB University Press, October 2001). Available at: https://www.marcprensky.com/writing/Prensky%20-%20Digital%20Natives,%20Digital%20Immigrants%20-%20Part1.pdf (accessed on 27 July 2018).

connection on a home computer with a minimum three years of Internet experience, concluded that[5]

> The internet provides an environment where adolescents can observe and learn attitudes and behavior, not only through frequent virtual interaction with known friends, both local and overseas, but also with global communities. The internet can be viewed as a virtual social system that allows adolescents to share their interests, express their opinions and form relationships and communities globally. These communities are drawn together because of mutual interest, and are not constrained by geography. This would imply that the socialization of teenagers is not restricted to the traditional sources of family, friends, school and exposure to passive media such as television. Rather, the active interaction allowed by the internet means attitudes may now encompass a global view.

The world of digital since then has grown even more ubiquitous. Greater numbers of people have taken to the digital landscape, and with greater frequency. The socialization process being engineered globally by a digital universe is seeping into consumer territory too.

DIGITAL CONSUMER SOCIALIZATION

Consumer socialization is seen as the process by which young people pick information and skills, and learn and harbour attitudes that allow them to operate as consumers in a marketplace. Until the digital age, the agents of consumer socialization were restricted to family and peers and those that belonged to a particular class and culture. Marketers too with clever use of media platforms joined in to influence and exact sought-after consumer behaviour from those growing up. As with socialization, the world of digital has radically altered the process of consumer socialization. Children growing up,

[5] Lee, Christina K. C., Conroy, Denise M. and Hii, Cecilia. 'The Internet: A Consumer Socialization Agent for Teenagers', ANZMAC 2003 Conference Proceedings Adelaide, 1-3 December 2003.

though exposed to consumption practices at home and within the community, are fast picking consumption attitudes in parallel from digital influences that seep in from around the world. To understand and participate in the digital consumer socialization process, business and brands have to first unravel and understand digital social behaviour. The mistake brands may make is to assume that digital socialization and digital consumer socialization automatically mean that digital natives will allow for brands to participate and engage in their digital lives.

A Forrester study conducted on 12- to 17-year-olds in the United States from half a decade ago is an eye-opener in this regard. Here's some of what the study found. Half of the youngsters surveyed didn't think too much of brand using social media. In fact, only 6 per cent from among them wanted to be friends with brands on social media networking sites. Think about that. The data mean a whopping 9 out of 10 youngsters don't want to make friends with brands on social media. A measly 16 per cent of young consumers surveyed expected brands to use social media tools to engage with them, and 28 per cent required that brands listen to what they had to say and respond if they had queries. Now contrast this with what the youngsters were actively doing on social media. A substantial three out four young people aged 12–17 were using social media platforms to talk to their friends about products they were interested in. Within a span of a year of the study, data revealed that the number of youngsters who posted reviews and ratings on online sites had doubled. Considering that 12- to 17-year-olds were more interested in engaging with each other rather than brands on social media, it is important that marketers listen and learn from these conversations. Also, when an opportunity presents itself to engage, say via a rating or review posted by a digital user, it's important for brands to seek greater insights and feedback via social media.

The Forrester study is a significant pointer to how business and brands can engage with younger audiences on social media.

However, the larger question still remains about how brands can become part of the digital consumer socialization process. The study by Christina, Denise and Cecilia can be the starting point to accomplishing this. Note that their study established that digital users learn differently in the socialization process from their predecessors who had no access or exposure to digital agents, content, devices, apps and technology. The study found that adolescents aged between 13 years and 19 years who lived digital lives learned from virtual word of mouth. Their purchase choices and attitudes were engineered via recommendations from their virtual social circles. There was also mutual learning where the digital group learned from each other, influencing one another on values, beliefs and/or behaviour. To understand why this happens in the digital world with far greater frequency and impact, you have to contrast learning that happens within the home offline with that which happens online. There are two critical differences to take note of. The first is that offline learning that happens within traditional physical environments flows via a hierarchy and is mostly top–down. A common example would be the learning passed on from the parent to a child. Hierarchical top–down flow of information and order has always had its disadvantages and limitations in engineering active engaged learning. Contrast this with virtual peer social networks. There is no hierarchy, and the flow is even-sided when it comes to power equations (at least in comparison to formal learning in physical spaces like the home). The learning is mutual and not one-sided, and therefore far more impactful. The second differentiator is that digital learning that progresses the socialization process isn't limited by geography or participation. In fact, such learning has an infinite potential and, by its very nature and structure, encourages back-and-forth and many-to-many engagements. It allows for anonymity if sought after by participants, as much as it allows for open-declared participation. In effect, digital social learning has no boundaries whatsoever.

In addition to word-of-mouth and mutual learning and influence, the research study by Christina, Denise and Cecilia found that 13- to 19-year-old adolescents used construction and discovery to understand and learn. It means that there was rampant use of the trial-and-error technique and learning as gained through 'experimentation'. Here is what one of the participants in the study said:

> I learn about the other bits and pieces, like searching, down-loading, and stuff by myself. I have the habit of exploring by myself and try things out…. I just click on anything that I felt might be relevant in my searching process and learn as I experiment new things.[6]

Further, the study found that the pursuit of multiple activities at the same time in the digital domain, a form of learning termed bricolage, meant that digital citizens were organizing inform-ation and engagements in a customized manner to suit their own individual styles, thinking and learning. In traditional, physical learning environments, this would be almost impossible. The final element revealed by the study showed that participants got their information from multiple sources which in turn indicated how flexible learners were and how open they were to varied viewpoints in their learning journey.

Although the process of socialization continues through an individual's lifetime, the early years are the most critical ones as normative social influences that prompt conformity are at play during that time. Thus, it becomes important for marketers to know how they can participate and be among the set of influencers shaping young psyches. Note that this must be done with utmost care using digital media, as any 'push' to get in brand messaging will be met with a 'pushback'. Here's one way to get this right.

---

[6] Lee, Conroy and Hii. The Internet.

Rather than relay messages to a young target audience, it's better to be found by them when they search for what they need. Here's an example of how this can be done. My daughter Brooklyn loves to do DIY (Do It Yourself) stuff at home. To get ideas on her craft, she scouts YouTube videos on her personal tablet. I asked her which are her favourite DIY channels, and she told me that they are Troom Troom, SaraBeautyCorner, and LaurDIY. Brooklyn told me that 'SaraBeautyCorner' is more about 'lifehacks', and that she loves getting advice on how things in her life can be done better and easier. A search on Google tells me that a 'lifehack' is 'a strategy or technique adopted in order to manage one's time and daily activities in a more efficient way'. Here is an opportunity for brands to become part of a daughter's daily life. It's painfully obvious that no brands are tapping in and that the space to influence my daughter is being occupied by private individuals who have now become YouTube stars. The data[7] on this is revealing. Just over a year ago, 42 of the top 100 YouTube channels were targeting children. In March 2016, these 42 channels generated among them a whopping 10.3 billion video views. Among them, only two were run by well-known traditional children-targeting brands. They were Disney Junior, UK's channel, which came at number 50 on the Tubefilter chart with 200.7 million views, and LEGO, which came at number 63 with 169.9 million views. Compare the views Disney and LEGO together have been able to garner with what Ryan ToysReview, hosted by seven-year-old Ryan, gets on YouTube (the difference in favour of the latter is 275 million views) and you'll realize how brands are almost clueless about the digital bus.

What a wasted opportunity to have become part of the socialization scene that kids like my daughter immerse themselves in. It's

---

[7] Stuart Dredge, 'How Toy Unboxing Channels Became YouTube's Real Stars', *The Guardian*, 28 April 2016. Available at: https://www.theguardian.com/technology/2016/apr/28/children-toys-unboxing-channels-youtube-real-stars (accessed on 27 July 2018).

Table 7.4 Audience Composition by Platform (Q2, 2017)

Platform	Adults 18+ (%)	P18-34 (%)	P35-49 (%)	P50+ (%)
Live + DVR/Time Shifted TV	41	26	34	52
AM/FM Radio	17	17	17	17
TV-connected Devices	6	12	6	3
Internet on a Computer	8	8	10	7
App/Web on a Smartphone	23	31	26	17
App/Web on a Tablet	5	6	7	14

Source: Nielsen Insights.

important to note that the process of socialization is a lifelong one. So when brands look for opportunities to be part of the process, they must do so keeping in mind their target audiences. To participate in the social scene their audiences take to, they must first decipher what their targets' interests are, where their digital selves are and how their targets socialize in the virtual world. *The Nielsen Comparable Metrics Report*, for example, reveals content consumption habits across age groups (Table 7.4). The study averaged across Americans tracks content consumption on TV, radio, TV-connected devices, PCs, smartphones and tablets. Optimizing content for these devices needs to be part of a socialization strategy pursued by brands.

DIGITAL ACCESS TO SOCIALIZATION

The key to understanding this opportunity that brands have to subtly teach consumers via the socialization process is due to the unhindered access they have to their audiences via digital platforms. In the world of physical, there is far less possibility of getting through. Also note that the access digital provides is infinite in that it isn't limited to a particular life cycle stage. Digital socialization can happen for a lifetime. Take a moment to imagine such a lifetime opportunity. Its potential is immense. The data on digital social time spent by adults are encouraging. An eMarketer report[8] of 2017 reveals that the prime choice of a social network remains FB, where American adults spent an average of 25 minutes per day. It's predicted that this time spent will increase by two minutes in the year 2019. Instagram, close on the heels of FB, is currently used for around 8 minutes per day, which will go up to 10 by 2019. Snapchat averages at five minutes per day, which will rise to eight minutes in 2019. All the other social channels put together including LinkedIn, Twitter and Pinterest take up 11 minutes every day.

The challenge for business and brands is to participate with their target audiences without sticking out like sore thumbs on virtual social platforms or in the digital world. The ones breaking through with relationship-oriented connections with their audiences are building devices and integrating them with digital technology platforms in a manner where they are turning into daily habits. Here's how Amazon and Google are doing this via their smart speakers Echo and Home and the built-in voice-controlled intelligent personal assistant service. Data released by Google in 2018[9] show how quickly these have become part and parcel of people's lives. One a personal note, I can vouch for this considering my family makes it a point to talk to Alexa every day. The lady of the home is getting Alexa to make 'to-do' and 'shopping lists'. My kids are asking to know about stuff and even talking to Alexa, though wishes of good morning and evening. I too ask when I need to know about something connected to schedules and other plans. To tell you the truth, Alexa is almost another member in our home. The Google data back up what my family feels. According to

[8] eMarketer, 'Adults Spend Half of Daily Media Usage on Digital', Inside Radio, 10 October 2017. Available at: http://www.insideradio.com/free/emarketer-adults-spend-half-of-daily-media-usage-on-digital/article_24addc46-ad97-11e7-bbda-4f926ba26027.html (accessed on 27 July 2018).

[9] Daniel Terdiman, 'Here's How People Say Google Home and Alexa Impact Their Lives', Fast Company, 1 May 2018. Available at: https://www.fastcompany.com/40513721/heres-how-people-say-google-home-and-alexa-impact-their-lives (accessed on 27 July 2018).

the study report that surveyed over 1,600 voice assistant owners, 41 per cent of those surveyed said that they felt they were talking to a friend or at least another person while engaging with the voice assistants. The report stated that users were talking back to voice assistants with words such as 'please', 'thank you' and even 'sorry'. It seems like people don't think it is a device but a person who is a friend, or as in our case, Alexa is seen as a family member. Seventy-two per cent of those surveyed for the study stated that the devices had become part of their daily routines, helping them plan for their day. Where were they placing these devices in their homes? Fifty-two per cent put it in their bedrooms, and almost an equal split of the rest had the devices in their bedrooms and kitchens.

The data results seem remarkable in that within a short span of time, technology brands have made inroads into people's lives to the extent that they have now turned into daily habits. The younger generation in fact grows up with such devices. In effect, what Amazon and Google have done though their technology and devices is enabling them to know us intimately. This infinite 'intimate' knowledge allows such them to engage with us in long-term relationships that have 'lifetime' commercial value. This is something marketers could only dream of in the past. The latest from Amazon Alexa is its stepping into work territories. Late last year, Amazon released Alexa for Business that allows employees to use virtual assistants to set up meetings and calls, book conference spaces, schedule vacations and so on. With data analysis peaking, there's no telling what Alexa can do for business leaders and decision-makers. All in all, this illustrates how brands that leverage the power of digital can be leveraged to become part of people's everyday lives, inside and outside of their homes.

CHAPTER 8

Outing Everything

Even while taking my courses as part of her MBA programme, Priyanka exhibited a keen eye and grasp for fashion. It wasn't, therefore, surprising that post MBA, when she started on her corporate career, she balanced along a stint in the fashion industry as a model. This went on for a while for Priyanka until motherhood beckoned. Having found her hands full, she decided to drop her modelling career and stick to her day job in HR. Never wanting to completely abandon fashion, she took to the digital platform to connect with her followers. This time around, she decided to focus on mommy fashion. After all, there were many mothers who reached out to her for fitness, diet and fashion tips. To ensure she kept her audience interested, she started putting out her life as a mother on Instagram. Keeping her account private, she started attracting and accepting followers, many of whom were mothers who wanted her beauty and fitness secrets. When I asked what she expected Instagram to do for her, she replied,

> That's my story out there. My life, more so as a mother is out there for people to see. Though I am selective about who I accept as followers, I am keen on them realizing that looking gorgeous and fit doesn't need to stop after motherhood.

Having Priyanka as part of my Instagram network, I knew of her popularity among women, young and old. When I asked her how she decided on the pictures she put up, she told me about wanting them to portray her as the fashionable and fit mommy that she was. 'I put pictures about my time at social places and occasions including hotels, vacation spots, parties, night outs and so on. You can see many of my pictures are with my daughter, Norah'. 'Is that the "real" you', I asked. 'It is a part of the "real" me', came the reply.

When I lingered on the pictures on Priyanka's Instagram account and studied it closely, I could see what she was talking about. Her pictures were telling a story of her social life. The pictures were both intriguing and attractive at the same time. In fact, I have myself wondered how she could stay this fit post motherhood. That question seems to be on many minds judging by the queries she gets. All the questions she gets have prompted Priyanka to take time out to engage with her followers, especially moms who are eager to know more about her fitness regime.

Now it isn't just Priyanka who takes to Instagram to tell her story, it's millions more. Worldwide, according to data from 2017, there are 500 million users who share 5 million photos and register 4.2 billion 'likes' every day. India on its part has 38,000,000 users who use apps to share photos and other visual material (see Table 8.1).

For many on Instagram, the visual content sharing that happens may not necessarily be a peek into their private lives but is certainly an outing of their public social lifestyles. A recent research study I conducted explored the display of virtual social personas and if they mirrored those being displayed in the real world. Further, I studied if persona exhibitions online effected through the use of lifestyle brands revealed a congruity between the public image the

Table 8.1 Indians on Instagram (2017)

Age	Users	Location	Users	Interests	Users
13–18	5,900,000	Mumbai	2,200,000	Bollywood	5,400,000
19–24	19,000,000	Delhi	2,700,000	Auto	19,100,000
25–35	10,000,000	Hyderabad	1,200,000	Music	28,300,000
36–45	1,900,000	Chennai	330,000	Travel	26,800,000
46+	1,100,000	Pune	2,300,000	Sports	29,300,000
		Ahmedabad	2,150,000	iOS	2,200,000
		Kolkata	1,260,000	Android	8,700,000

Source: SocioAdvocacy, 'Mind-blowing Stats About Indians on Instagram in 2017', SocioAdvocacy, 2 May 2017. Available at: https://www.socioadvocacy.com/blogs/instagram-audience-in-india-2017/ (accessed on 27 July 2018).

patronized brands sported and the aspirational selves sought after by those making the displays.

The conclusions I came to were remarkable to say the least.

DIGITAL SOCIAL PERSONAS

The term 'persona' is Latin in its origin and literally means 'mask' or a character played by an actor. The Swiss psychiatrist Carl Jung characterized persona as the 'social face' people put on to make particular impressions on others, and also to keep in check their true natures. Although personas help in establishing social images and relationships, Jung did caution on an 'over-identification' with one's persona[1]:

> Fundamentally the persona is nothing real: it is a compromise between the individual and society as to what a man should appear to be. He takes a name, earns a title, represents an office, he is this or that. In a certain sense all this is real, yet in relation to the essential individuality of the person concerned it is only a secondary reality, a product of compromise, in making which others often have a greater share than he. The persona is a semblance, a two-dimensional reality.

Jung may never have foreseen a digital world where persona exhibitions play out in plenty. My study on the display of digital social personas revealed insights that have major implications for business and marketing.

Last year, the *Daily Mail* in the United Kingdom ran two stories that had the terms 'Luxury Kids of Instagram'[2] and 'Rich Parents of

---

[1] http://www.psychceu.com/jung/sharplexicon.html (accessed on 11 September 2018).

[2] Sofia Brennen, 'Are These the Most Outrageous Displays of Teenage Wealth Yet? Luxury Kids of Instagram Flaunt Their VERY Lavish Lifestyles—Including Pouring Champagne over Cereal and a Pet Tiger', *The Daily Mail*, 4 October 2017. Available at: http://www.dailymail.co.uk/femail/article-4944804/Luxury-Kids-Instagram-flaunt-lavish-lifestyles.html (accessed on 27 July 2018).

Instagram'[3] in their titles. The first article featured rich kids who put up pictures that featured in them the brand they patronized. These brands included Christmas Fantasy spring water that costs €300 a bottle, a Louis Vuitton blanket, Ace of Spades champagne that's priced at £595 for a magnum, Versace clothing and accessories, and luxury cars. The moneyed parents on their part put up pictures that featured luxury brands and products such as Rolls-Royce, Lamborghini, Armand de Brignac champagne, Louboutin trainers, Louis Vuitton bowler bags and even private jets. It isn't just the moneyed class that's showing off the brands they own on social media but even the classes below. The latter display for public consumption the brands that they can afford to own and possess. It isn't just product brands that people feature in the content they post on social media, but it includes service brands too. People in the middle classes put up social posts and pictures that feature cars, motorbikes, electronic gadgets, restaurants, hotels and even exotic pets. What all of them are doing both consciously and unconsciously is constructing and displaying aspirational personas for social consumption. Also, this is being done across multiple social media platforms. In fact, data from the United States clearly show people as exhibiting substantial reciprocity across social media platforms, meaning they are on and use multiple social apps at the same time (Table 8.2).

Although each of these social media platforms has its unique features tailor-made for its target audiences, the uniting factor is that they give their users opportunities to create and display content for social consumption. This digital content put out in social settings is valuable to the extent that it incisively reveals the exhibitor's psyche and also what the person in question intends to

---

[3] Sofia Brennen, 'The (VERY Wealthy) Bank of Mum and Dad: Rich Parents of Instagram Put Their Children in the Shade with Their Boastful Photos of Pet Cheetahs, Private Jets and Wads of Cash', *The Daily Mail*, 7 June 2016. Available at: http://www.dailymail.co.uk/femail/article-3629123/Rich-parents-Instagram-lavish-lifestyles.html (accessed on 27 July 2018).

Table 8.2 Parallel Use of Social Media Platforms (Percentage of Users Who Also...)

% of Users	Use Twitter	Use Instagram	Use FB	Use Snapchat	Use YouTube	Use WhatsApp	Use Pinterest	Use LinkedIn
Twitter	–	73	90	54	95	35	49	50
Instagram	50	–	91	60	95	35	47	41
FB	32	47	–	35	87	27	37	33
Snapchat	48	77	89	–	95	33	44	37
YouTube	31	45	81	35	–	28	36	32
WhatsApp	38	55	85	40	92	–	33	40
Pinterest	41	56	89	41	92	25	–	42
LinkedIn	47	57	90	40	94	35	49	–

Source: Pew Internet, 'Substantial "Reciprocity" Across Major Social Media Platforms', Pew Research Center, 27 February 2018. Available at: http://www.pewinternet.org/2018/03/01/social-media-use-in-2018/pi_2018-03-01_social-media_0-04/ (accessed on 27 July 2018).

construct and display as a social image. The multiple use of social media platforms is to construct these images keeping in the mind a particular social context. My research comprehensively revealed that people construct social images for public consumption in the real world and on the digital realm, and that the two mirror each other. The digital platform provides far greater opportunities for image construction and propagation in comparison to what can be achieved in the real world. These 'infinite' opportunities mean that people can construct and display contrived personal images to an unrestricted audience, if need be. Also, there is no limitation in terms of the quantum of content and until what time.

At the heart of the need for a social image is the craving for what is termed the 'ideal social self'. Research proves that part of the way we see ourselves is fashioned by the way others see us. The concept of the self and our sense of esteem about ourselves is derived through social comparisons. There is also self-presentation we do within our social settings wherein a positive self-image is conveyed with the goal of upping our social status. The personality traits possessed by an individual have a strong influence on the way a self-image is constructed and conveyed. In the virtual world, for example, a study conducted in Taiwan by Shaojung Sharon Wang and Michael A. Stefanone in 2013[4] explored how the functioning of personality traits such as extraversion and narcissism influenced self-disclosure which in turn impacted the intensity of 'check-ins' on FB. Although the study concluded that the listed traits may not have directly influenced 'check-in' intensity, it did reveal that

> The indirect effects of self-disclosure and exhibitionism were particularly salient in predicting the intensity of check-ins. The study found that self-disclosure not only had a significant direct effect on Facebook check-in intensity but also had a significant indirect effect on check-in intensity through exhibitionism. There was also a significant direct effect from extroversion to self-disclosure that contributed to a complete path from extroversion to Facebook check-in, through self-disclosure and exhibitionism. Moreover, although there was no direct effect of narcissism on Facebook self-disclosure, a path from narcissism to check-in intensity through exhibitionism was discovered.

It isn't just this study that confirmed links between personality traits and social media content construction and propagation.

DIGITAL PERSONA CONGRUENCY

Half a decade ago, a study[5] from the University of Illinois that examined the strength of consumer–brand relationships revealed

[4] Shaojung Sharon Wang and Michael A. Stefanone, 'Showing Off? Human Mobility and the Interplay of Traits, Self-disclosure, and Facebook Check-ins', *Social Science Computer Review* 31, no. 4 (2013). Available at: http://citeseerx.ist.psu.edu/viewdoc/download?doi=10.1.1.723.4896&rep=rep1&type=pdf (accessed on 27 July 2018).

[5] Shirley Y. Y. Cheng, Tiffany Barnett White, and Lan Nguyen Chaplin, 'The Effects of Self-brand Connections on Responses to Brand Failure: A New Look at the Consumer–Brand Relationship', *Journal of Consumer Psychology* (15 June 2011). Available at: https://onlinelibrary.wiley.com/doi/pdf/10.1016/j.jcps.2011.05.005 (accessed on 27 July 2018).

that those individuals who had high 'self-brand connections' responded to negative brand information as they did to personal failure. That is, they perceived such failures as being a threat of their positive self-view. The study further found that such individuals were reluctant to lower brand evaluation despite negative information about such brands. The reason for such reluctance was a desire to protect their own sense of self-esteem rather than the brands in question. There are umpteen research studies out there that prove that people patronize those brand that allow them to present either their actual or their ideal selves for social consumption. A study[6] from the University of Georgia showed that there are times when consumers use brands to reflect both their actual and ideal selves on a social media platform like FB. With the ubiquitous use of digital media platforms by users, self-expressions by those who inhabit such virtual social spaces have ballooned to unimagined proportions. All such digital displays of the self by users are invaluable material for business and marketing.

Without transgressing into invasion of privacy territories, it is important for marketers to use digital content that reveals the self to make better marketing and branding decisions. What is common practice today is tracking, collecting and analysis of digital footprints left behind by those in the digital world. Much of such collections and analyses focus on consumption journeys and decision-making. I am recommending an addition to such data collection and analysis, to be effected in the domain of digital personas and lifestyles. In research parlance, that would mean adding to the collection of what is currently consumption data. It isn't enough that digital data be mined to know how buyers are

---

[6] Candice R. Hollenbeck and Andrew M. Kaikat, 'Consumers' Use of Brands to Reflect Their Actual and Ideal Selves on Facebook', *International Journal of Research in Marketing* (16 September 2012). Available at: https://media.terry.uga.edu/socrates/contact/documents/2016/02/29/Consumers_use_of_brands_to_reflect_their_actual_and_ideal_selves_on_Facebook.pdf (accessed on 27 July 2018).

making consumption decisions online. The digital lives of people allow for more than a peep into their psyches, motivations and aspirations. This is rich data that can be leveraged for say brand personality decisions. Such digital content data is in fact more revealing than psychological research conducted in real-world physical settings. When consumers are probed in the proctored physicals settings, there is a tendency to present to the researcher what are apt responses. Although qualitative psychological research methods are used in such settings, there is still no guarantee that what is being revealed reflects the 'truth' about consumer motivations and desires. Digital social content constructed and propagated every minute by those who live parallel existences on virtual media is the answer to 'getting to the truth'. The unbridled construct and relay of aspirational social images in the digitized social world is a gold mine that marketers can tap into to engage and connect their brands with their target set of consumers. Also note that such digital access to social image building has no limitations of either geography or time. Meaning, business and brands can access such content put up for display anywhere in the world, and over the lifetime of their buyers, assuming that content is generated for such periods.

A pertinent example that can illustrate the use of social media content would be 'slice of life' advertising. The latter is a technique used by marketing communicators where a real-life problem scenario is presented in an advertisement or commercial and the brand advertised is presented as the solution to the problem. Digital social content is rife with depictions of real-life scenarios. When Bitstrips was around, social media used to be inundated with people putting up events about their everyday lives in the form of cartoon strips. Such 'slice of life' display is almost a dream come true for brands trying to take a peek into people's private lives. Although the comic strip series isn't around anymore, Bitstrips developed and now operates 'Bitmoji', a mobile app that enables users to build and display personalized avatars in their messages to one another. Instagram is another app that abounds with displays of people's lives. When the photo- and video-sharing social networking service allows for stories in the form of pictures and videos to be put, the material displayed is again a peek into people's everyday lives. Such 'peeks' can enable marketers to know about people's food habits, pet loves, sports club fandom, travel diaries, fitness achievements and so on. A series of picture stories on Instagram can reveal patterns from everyday lives. Imagine if text, image, audio and video content from across social media platforms is collected and analysed for a target audience, what it may reveal may never be matched by what 'real-world' research can. One such virtual study that I undertook combined the technique of content analysis with phenomenological analysis. I conducted a series of 'real-world' depth interviews over a period of a year with a group of college-goers in the city of Bengaluru. The data I collected through the interviews and subsequent phenomenological analysis revealed a series of image constructs the college-goers tried to build and display in their social settings. I termed one of such image-building displays as exhibitions of a 'balanced' lifestyle. A significant percentage of the college-goers I studied included those who had to come to the city of Bengaluru for higher studies from smaller cities and towns across the country. These students came from conservative families, and so had ingrained habits and behavioural traits that prompted them to operate within the dictated social norms. The city of Bengaluru and its cosmopolitan lifestyle enticed them to abandon their conservative ways, which they did but only within certain limits they had set for themselves. So, for example, going to parties was okay but getting into a relationship wasn't. Having a good time at the dance floor was permitted but getting drunk wasn't. Note that these were self-imposed restrictions that tried to balance between the need to stay conservative and tug of liberal indulgences. When I studied, with due permission, the virtual social content these youngsters displayed, I found in them the display of what I had termed a 'balanced lifestyle'. The photos, for example, the college-goers put up and the issues they wrote about did reveal their lives to their social audience, but it did so in a manner where they were seen as having a good time without any crossing of established 'limits'. One example of how this was achieved was through putting up 'cropped photos' that cut anything that would be deemed inappropriate on Instagram or FB. The youngsters in question consciously ensured that the required edits were done before content went up on virtual social display as they knew this was also being viewed back home. It was, therefore, important that 'socially responsible' content only be seen on public platforms. The identification of such 'socially responsible funsters' was an outcome of research engagements that I conducted in person and through the study of digital content. I am convinced that going forward, the 'face-to-face' can be eliminated and replaced with virtual social content analysis. In fact, what the latter can reveal about buyers and their lives, the former just cannot.

In the industry too, offline and online research are being used to identify and characterize users based on their social networking inclinations. A 2014[7] study by Coca-Cola Retailing Research Council of North America and the Integer Group using a quantitative study first identified four distinct social networking personas and uncovered their mindsets, motivations and behaviours. The identified groups were christened 'Bonders', 'Sharers', 'Professionals' and 'Creators' (Table 8.3). This was followed up by a qualitative research study that uncovered each of the groups' attitudes and behaviours as they relate to social networking and shopping. The study found that Bonders were relationship oriented and connected well with their family, friends and colleagues. They saw themselves as the fun-loving types who hosted and planned local

---

[7] Integer and Coca-Cola Retailing Research Council, 'Untangling the Social Web: Insights for Users, Brands, and Retailers', Coca-Cola Retailing Research Council, March 2012. Available at: http://www.ccrrc.org/wp-content/uploads/sites/24/2014/02/Untangling-the-Social-Web_Part-31.pdf (accessed on 27 July 2018).

Table 8.3 Time Spent on Social Media

Social Persona	Every Other Day or So (%)	Less Than an Hour (%)	1–2 Hours (%)	3–4 Hours (%)
Creators	8	0	18	7
Bonders	16	28	25	31
Professionals	43	47	28	40
Sharers	33	25	20	23

Source: Integer/iModerate, *Social Networking and Brand Engagement Research*, 2011.

events. Their desire to socialize came from an interest in knowing and catching on the lives of people in their social circles. The advent of virtual social networking shifted the Bonders on to sites like FB. Bonders stay longest on FB and use it extensively to connect and catch on other people's lives. Such an indulgence is among the favourite pastime activities of Bonders. The group identified as 'Sharers' loved to spread the word on what was going on in their lives, and also information that they believed was helpful to others in their social circle. The driver for Sharers was the need to build strong relationships with others through helpful acts. Sharers were kind, helpful and sincere people who felt a sense of 'usefulness' in doing the good acts of helping others. They too depended on FB for much of their social acts. The 'Professionals' consisted of people who were focused on their careers and used the social web for networking and professional advancement. This group used blogs, videos and business networking sites like LinkedIn along with FB to pursue their networking objectives. The Professionals were introverted and were keen on being perceived as intelligent, efficient and organized. The 'Creators' aggregated as the bold, creative, outgoing types who were keen on building and sharing original content on the social web. Most of the Creators were active on FB and Twitter and were posting content frequently while watching videos, reading blogs and browsing the social web for updates on their dear ones.

LEVERAGING PERSONAS

The identification of personas by tracking users' socializing habits is useful to business and brands to the extent of knowing how to engage with them using the right digital tools and platforms. Going a step further would mean that marketers use public social content (with due permissions) to understand people's lives. So when Parul uses Bitstrips to put up a cartoon that shows two unhappy girls on the phone with a strapline, 'Parul and Gunjan are terrible at making plans', they are letting their social circle in on how they mess up at times. When Raghav put his Bitstrip up, this is what is comically communicated via a cartoon, 'Happy Birthday, Mama!' The strip shows his mama opening her presents with jubilation all over her face. Raghav standing next to his mom is sporting a big grin. He looks pleased. Below the picture is the reply from Raghav's mom, 'Thanks Raghav! Love the presents! Thanks, beta (son)'. It's easy to see from the cartoon strip that Raghav dotes on his mom. His mother's birthday means a lot to him. What about Parul and Gunjan? We now know that they like to hang out together, and that they get their plans messed up at times. The clothes they are wearing in the cartoon suggest that it's probably a nightspot they were planning to go to. They could in all probability be party animals. Their virtual social content, if studied carefully, can confirm that.

The advent of digital has enabled people to put their lives into the open. Such virtual displays, in addition to being material for communication, are also acts aimed at image building. When digital wasn't around, impression management was mostly restricted to people using their selves to construct and display aspirational identities and images. With digital, such impression management has stepped into territories that include homes and other private spaces. People online aren't just showing themselves off but are also making statements with their homes, possessions, gadgets and even their kids and pets. In a study[8] by Gemma Kennedy and Dr Elvira Bolat at the Bournemouth University on conspicuous luxury consumption in the social media context that centred on HENRYs (high earners, not rich yet), it was found that this high-income group purchased luxury for the purpose of displaying status. Further, it was found that in the context of social media, it became even more pertinent to demonstrate luxury possessions HENRYs owned through displays of social media content. The study reported, 'HENRYs aspire to purchase luxury by mostly depending on user-generated word-of-mouth and are driven to produce social media content as evidence of luxury purchasing and possessions and to satisfy narcissistic ambitions'.

The desire to display possessions isn't just restricted to products and services in the luxury category. People desire to show off possessions they value. So in India, for example, a middle-income family that moves 'up' from a two-wheeler to buy their first car would treat the vehicle as a status symbol within their social class. Among a class of two-wheeler owners, a car is quite a status symbol. So is a smartphone, or a smart TV. People who own such products treat them as status symbols within their community. What digital enables such people to do is to take displays of such status symbols far and wide. So when Arpan, a migrant guard working with a security agency in Bengaluru, puts up a picture of him talking on his smartphone on FB for people in his social circle that include those in his home state to see, it is a display of status within a certain social class. Such displays have been made possible by social media. In the olden days, it would have taken Arpan a travel back to his hometown for showing the smartphone off. In an era of digital, Arpan can convey his status message literally without moving an inch.

---

[8] Gemma Kennedy and Elvira Bolat, 'Meet the HENRYs: A Hybrid Focus Group Study of Conspicuous Luxury Consumption in the Social Media Context–Competitive Paper', 2017. Available at: http://eprints.bournemouth.ac.uk/29423/3/AM17_0335_competitive.pdf (accessed on 27 July 2018).

The implications of digital image displays are big on business and marketing. What digital has done is turned everything about a person's life into broadcast material. The impetus to broadcast products and services comes from their value as a 'show-off' material. Note that such displays are a global phenomenon, afflicting anyone and everyone across social classes. Social status displays have an infinite reach. They can travel to anywhere at any time. Smart businesses are those which have realized the potential that this desire for digital display holds. They have understood that conspicuous consumption isn't just a luxury category phenomenon; instead, it's an every category possibility. Anything that can be displayed, and thanks to digital anything can, is material for digital image building and propagation. If people's homes feature in pictures on social media, you can bet they would want their places of habitation to look aesthetic and classy. It's one of the reasons why the home decor market has rocketed in India. According to industry estimates,[9] the Indian home decor market is growing 15–20 per cent a year and it includes buyers from small cities and towns. Last year, it was reported, 'When Karan Johar posted a picture of his redesigned terrace, he left 3.9 million Instagram followers with an overwhelming desire to give their own homes a makeover, draped in luxury'. Did someone actually do a makeover of their home? Here's what the *Economic Times* reported:

In deepest Andheri, along Mumbai's western water margin at Versova, Subadra Pillai followed the filmmaker in giving her 15-year-old apartment a facelift that would blend style with understated elegance. The 44-year-old architect, who had hitherto expended scant little on the apartment, recently spent Rupees 60 lakh in three months to paint her imagination onto

---

[9] Bhagyashree Nair and Richa Maheshwari, 'Demonetisation-hit Luxury Home Decor Business Rebounds to New Highs', *The Economic Times*, 29 April 2017. Available at: https://economictimes.indiatimes.com/magazines/panache/demonetisation-hit-luxury-home-decor-business-rebounds-to-new-highs/articleshow/58424997.cms (accessed on 27 July 2018).

the 1300 square-feet canvas. The expenditure amounted to a fourth of the basic cost of a property in the tony western suburbs in India's commercial capital: The plain whitewashed house gave way to exquisite wall coverings, old home décor was replaced with Persian rugs, and sequined fabric boxes and delicate silver-plated vases were brought in to lend character and identity to the home.

Indian home decor brands have been using digital content and platforms in the past few years to reach their target audiences, both inside and outside the country. What decor buyers on their part have been doing is using digital content to paint pictures about their lives. Such digital pictures have also done their part doubling as status displays. It isn't just the business of decor that's aiding consumers in status displays; it's almost anything and everything they own. A kitchen gadget, bedroom furnishing, living room furniture or even the powder room fitting is now open to public social display, thanks to the world of digital. It in turn means that marketers aren't just designing and building goods and services for utilitarian purposes and that the aesthetics and display value of everything they make is critical. After all, consumers out there if given an opportunity to digitally pirouette their possessions will do so. Tapping into the potential for display that digital affords buyers with possessions is an opportunity that will remain, as long as digital is part and parcel of people's lives.

CHAPTER 9

Inhibitions on the Loose

During the combined Senate Judiciary and Commerce Committee hearing investigating FB's abuse of its users' private data, this is what Senator Dick Durbin asked the FB founder–CEO, 'Mr Zuckerberg, would you be comfortable sharing with us the name of the hotel you stayed in last night?' The CEO's response after an awkward bit of silence was a 'no'. The Senator continued, 'If you've messaged anyone this week, would you share with us the names of the people you've messaged?' Again, the response was, 'Senator, no, I would probably not choose to do that publicly here'. The Senator then proceeded to put into perspective the reason for the hearing, 'I think that might be what this is all about. Your right to privacy, the limits of your right to privacy, and how much you'd give away in modern America'.

The Senator had a point.

America and the rest of the world, without realizing, are either collecting or giving way too much information away in the digital world. Consider this for a moment. According to the new 2018 Global Digital suite of reports[1] from We Are Social and Hootsuite, there are now more than 4 billion people around the world using the Internet. Out of that, a quarter of a billion new users joined the digital world in 2017 alone. With a global population of 7.593 billion and urbanization standing at 55 per cent, here are the digital use numbers. There are 4.021 billion people who use the Internet (53% penetration), 3.196 billion active social media users (42% penetration), 5.135 billion unique mobile users (68% penetration) and 2.958 billion active mobile social media users (39% penetration).

[1] Simon Kemp, 'Digital in 2018: World's Internet Users Pass the 4 Billion Mark', We Are Social, 30 January 2018. Available at: https://wearesocial.com/blog/2018/01/global-digital-report-2018 (accessed on 30 July 2018).

Since last year (April 2017), Internet users have grown by 7 per cent, active social media users by 13 per cent, unique mobile users by 2 per cent and active mobile social media users by 14 per cent. If you were to look at FB's numbers in April 2018, the social networking site has 2.234 billion active users every month. Since April of last year, FB population has grown by 14 per cent with 89 per cent accessing the site via their mobile phones. Of the declared profiles, the male–female split is almost even with 43 per cent declared female profiles and 57 per cent declared male profiles. Never mind privacy concerns in the wake of Cambridge Analytica, FB's user growth has climbed steadily. Between the months of January and March of 2018, 10 million Americans started using FB. That was an MAU (monthly active users) increase of 4 per cent. For the same period, in the United Kingdom, the number of users went up by a million (2% increase). In the Asia-Pacific region, FB's growth was even more impressive. For a period of three months starting January 2018, India added 20 million new FB accounts (8% increase) and Indonesia upped its numbers by 10 million (8% increase) (Table 9.1). These growing numbers indicate that the social content being either generated or tracked is on the upswing.

Table 9.1 FB Rankings and User Base

Rank	Country	Users	YoY (%)	QoQ (%)
1	India	270,000,000	27	8
2	United States	240,000,000	10	4
3	Indonesia	140,000,000	26	8
4	Brazil	130,000,000	6	0
5	Mexico	85,000,000	12	2
6	Philippines	69,000,000	10	3
7	Vietnam	58,000,000	16	5
8	Thailand	52,000,000	11	2
9	Turkey	52,000,000	8	2
10	United Kingdom	45,000,000	7	2

Source: Hootsuite/We Are Social, 2018.

DIGITAL PRIVACY PARADOX

A 2016 study[2] titled 'Privacy and Social Media: Do Users Really Care?' by Hannah Ersdal and Sølvi Svendby of the Norwegian University of Science and Technology explored the issue of privacy on social media platforms and if users knew and cared about personal information being collected and disseminated. The Norwegian participants (social media users) in the study ranged from 13- to 80-year-olds. Unlike what the researchers had thought at the beginning of the study, the level of education had virtually no impact on the participants' knowledge on issues of online information spreads. They had no much understanding of digital information sharing and although the majority among the respondents knew that they were being tracked, they had no idea about the presence of 'third parties' on portals they visited. Further, most respondents did not care to read and understand the privacy policy in place. They did not take the effort because they chose 'convenience' over the effort needed to know. The study found that older respondents were more sceptical in sharing personal information online as compared to the younger ones. The 'privacy paradox' the authors of the study arrived at is interesting to note. The study found that users claimed to care about issues of privacy online, and that they had enough information about digital information being collected and disseminated. However, they made no effort to read the documents being provided by firms online and utilize available services without a clear understanding of how technologies in place are collecting and using personal information.

More than half a decade ago, when European researchers looked at issues of 'self-disclosure' in a study[3] titled 'Online Social Networks:

---

[2] Hannah Ersdal and Sølvi Svendby, 'Privacy and Social Media: Do Users Really Care?' (Norwegian University of Science and Technology, 2016). Available at: https://brage.bibsys.no/xmlui/bitstream/handle/11250/2403229/15179_FULLTEXT.pdf?sequence=1 (accessed on 30 July 2018).

[3] Hanna Krasnova, Sarah Spiekermann, Ksenia Koroleva, and Thomas Hildebrand, 'Online Social Networks: Why We Disclose', *Journal of Information Technology* (June 2010). Available at: https://www.researchgate.net/publication/220220751_Online_Social_Networks_Why_We_Disclose (accessed on 30 July 2018).

Why We Disclose', what they found was that users were primarily motivated to disclose information due to the conveniences they enjoyed in maintaining and developing relationships and for platform enjoyment. The study empirically zeroed in on factors that influenced self-disclosure on online social networks. The benefits that digital platforms provided including convenience, relationship building and enjoyment were significantly found to influence information disclosure. Although the risks of such disclosure did put doubts in the minds of users, these were offset by benefits offered and mitigated by trust and control beliefs. Other studies have also identified factors such as peer pressure, perceived anonymity or anticipation of face-to-face encounters as contributors to digital self-disclosure. Although 'perceived privacy risk' had a significant negative impact on the information disclosed, users ensured that they adjusted the information being put out. They did this based on the perception of threats to privacy. Furthermore, the benefits gained from virtual secular platforms that enabled 'intensive communication' overrode the risks and induced greater revelation by users. The mitigating factors of 'perceived control' and 'trust' in the online social network provider ensured that risk perceptions were downgraded. Such control and trust were built partly out of functional features such as privacy settings and clear information on privacy-related procedures on networking portals.

It's one thing for social networking sites like FB to keep reams of user data, and another for users to have enabled the same. To go deeper and unravel why people actively parade to various degrees of private information about their lives in the digital world, it's important to understand how the medium by itself is an enabler. Professor John Suler's work[4] from more than a decade ago on 'online disinhibition' among others opens up on why people act

---

[4] John Suler, 'The Online Disinhibition Effect', *Cyberpsychology & Behaviour* (2004). Available at: https://pdfs.semanticscholar.org/c70a/ae3be9d370ca1520db5edb2b 326e3c2f91b0.pdf (accessed on 30 July 2018).

out the way they do when it comes to self-disclosure. Online disinhibition as a cyberspace phenomenon is about how people who inhabit the digital world open up to say and do the things they wouldn't otherwise say or do in the physical world. Such disinhibitions are classified as either 'benign' or 'toxic', and they work diametrically opposite to each other. Benign disinhibition happens when people share personal information about themselves, including talking about their wishes and fears. Benignly disinhibited individuals exhibit acts of unusual kindness and generosity, reaching out of their ways to help others. In contrast, those that act in toxic uninhibited manners in cyberspace indulge in harsh criticisms, display anger and hate, and make threats. Such people traverse the dark underbellies of the Internet accessing virtual places of pornography, crime and violence. Such toxic uninhibited behaviour exhibited in the digital realm is far from what such people practise in the real physical world.

The common thread across the two sets of behaviour is its nature of 'uninhibitedness'. John Suler's work posed the question of what causes such online disinhibition. Further, the elements of the digital world that contributed to the weakening of psychological barriers that block hidden feelings and needs were explored. Six factors were identified as being involved. For a few of those exhibiting disinhibition, one or two of factors contributed the larger share of influence; for most others, all the six factors intersected, interacted and even supplemented each other. This in turn ensured a complex and amplified effect. The six factors identified were dissociative anonymity, invisibility, asynchronicity, solipsistic introjection, dissociative imagination, and minimization of status and authority.

DIGITAL DISINHIBITION

In almost all my interactions with those who have slipped into the roles of trolls at some point in time in digital realm, the common characteristic that bound them together was their practice of 'dissociative anonymity'. The ability to practise their online lives differently from their real lives has contributed immensely to people taking to trolling. What surprised me was that a considerable number of those I spoke to did not even see their behaviour as bothersome. Would they have done the same in a physical setting? The majority responded with a 'no'. John Suler noted the same in his work on online disinhibition[5]:

> When people have the opportunity to separate their actions online from their in-person lifestyle and identity, they feel less vulnerable about self-disclosing and acting out. Whatever they say or do can't be directly linked to the rest of their lives. In a process of dissociation, they don't have to own their behaviour by acknowledging it within the full context of an integrated online/offline identity.

If anonymity is about concealment of identity, invisibility is about avoiding detection. The infinite digital world that's barrier-less presents perfect opportunity for navigation without the worry of social judgement. Invisibility amplifies on anonymity providing greater impetus to the practice of disinhibition.

The asynchronous nature of communication further enables the abandonment of inhibitions. In the real world, responses to what people say or do can be instantaneous. It's not so in the digital world. The courage to put out material is enhanced when there's greater opportunity to step away and delay reactions to what has been put out. Psychotherapist Kali Munro writing about the 'conflicts in cyberspace'[6] states,

---

[5] Ibid.

[6] Kali Munro, 'Conflict in Cyberspace: How to Resolve Conflict Online', Kalimunro.com, 2002. Available at: http://kalimunro.com/wp/articles-info/relationships/article (accessed on 30 July 2018).

You can say anything you think and feel without censorship at any time, including in the middle of the night when you're most tired and upset, leave immediately without waiting for a response, and possibly never return—in the extreme this can feel to some like an 'emotional hit and run'.

The limited stimuli that is available to netizens in comparison to what's encountered in the physical world means that people open up and express without inhibitions. In India, for example, when Virat Kohli, a famous cricketer, failed to score at a game, online trolls targeted his newly wedded wife and movie star, Anushka. That the trolls knew nothing about the lady didn't matter one bit. The trolling Anushka suffered prompted her husband to take to Instagram to make this statement[7]:

> Shame on those people who have been having a go at Anushka for the longest time and connecting every negative thing to her. Shame on those people calling themselves educated. Shame on blaming and making fun of her when she has no control over what I do with my sport. If anything she has only motivated and given me more positivity. This was long time coming. Shame on these people that hide and take a dig. And I don't need any respect for this post. Have some compassion and respect her. Think of how your sister or girlfriend or wife would feel if someone trolled them and very conveniently rubbished them in public. #nocompassion #nocommonsense.

Solipsistic introjection plays out when text messages or posts are read and the receivers perceive the reading as a voice in their heads.

People also have a tendency to engage in dissociative imagination in the digital realm. This is best illustrated when netizens adorn

---

[7] https://www.instagram.com/p/BDfY8nWh_HI/?hl=en (accessed on 11 September 2018).

online avatars when they operate in virtual environments. These avatars used, for example, in gaming then get extended into virtual social environments. Those who can't then dissociate the fantasies they experience from social realities will extend such fantasies into their virtual social environments. In the digital world, it's hard to keep up status hierarchies. It's the same with exhibitions of authority. In the world of physical, people who occupy the upper rungs of status and authority hierarchies rely on possessions, materials and even body language to convey their positions. The lack of such material for exhibition in the digital domain means that authority figures are hard to come by. The digital platform sans authority displays is an even keeled platform that diminishes the fear of disapproval and punishments. This in turn is greater encouragement for disinhibition.

MARKETPLACE DISINHIBITION

When Protein World unveiled its 2015 'Are You Beach Body Ready' campaign in the United Kingdom, the backlash[8] was both swift and relentless. Richard Staveley, the head of marketing for Protein World supplement sales firm, pointed to the brand's display being vandalized multiple times during their time at the London tube. Speaking to 'Good Morning Britain', he claimed to have received physical and violent threats at the company's head office, including even a bomb threat. A petition on Change.org received 60,144 signatures, asking for a takedown of the London Underground advertising. Despite the blowback, the company stayed the course with the campaign and claimed that the enhanced media exposure it received raked in £1 million in four days of sales. In fact, buoyed by the media time the brand received in the United Kingdom, the campaign was then taken across the Atlantic to New York City

---

[8] John McCarthy, 'How Social Media Told Protein World Where to Stick Its #beachbodyready Ads', The Drum, 29 April 2015. Available at: http://www.thedrum.com/news/2015/04/29/how-social-media-told-protein-world-where-stick-its-beachbodyready-ads (accessed on 30 July 2018).

where a huge billboard featuring swimsuit-clad model Renee Somerfield was unveiled at the Times Square.

Protein World may have been lucky to get away unscathed, but not so for the big daddy of burgers, McDonald's. When the burger giant unveiled its #McDStories campaign[9] on Twitter in 2012, the original idea was to have the brand's followers share their happy McMemories in 140 characters. The sharing did happen, only that there were more sour moments shared than happy ones. Shared Twitter stories ranged from 'Dude, I used to work at McDonald's. The #McDStories I could tell will raise your hair' to 'One time I walked into McDonald's and I could smell Type 2 diabetes floating in the air and I threw up'. Barely two hours later, the plug was pulled on the brand's Twitter marketing campaign.

The online disinhibition effect has had its trickle-down into the consumption arena. This has meant that the buyers who engage with brands online do so without the inhibitions that the physical marketplace would otherwise impose on them. Operating in an uninhibited manner doesn't have to be seen as a threat by business and brands. In fact, the benign form of disinhibition throws up a treasure trove of information spread that brands can leverage to both understand and build relationships with their target customers. Even if digital buyers step into territories of toxic disinhibition, there's much that they still reveal that can be useful to business and marketers. Here's an irate customer venting a mouthful via FB post at Indigo Airlines, a private carrier headquartered in India:

> Indigo flight, you suck! I land late night and the waiting period right from getting off from the flight to collecting luggage is taking more than the flight duration from Delhi to Bangalore. Pathetic service! No value for people's time. Easy approach.

---

[9] Kevin Fullerton, 'Hashtag Humiliation: The Tweets That Shamed a Brand', Attercopia, 19 April 2016. Available at: https://www.attercopia.co.uk/2016/04/19/hashtag-humiliation-the-tweets-that-shamed-a-brand/ (accessed on 30 July 2018).

Bloody unprofessional. All this after a hectic day and when I am sleep deprived. Indigo, get a little professional and stop snoozing when you have chosen to get into some real business with real customers. Indigo lost one customer today ☹☹☹☹☹!

This content was posted at 12.30 AM by the Indigo flyer. If the airline had reacted in real time, it could have engaged with the flyer on the social platform and outside of it to provide some relief. It could then have tried to make up if necessary for the less than satisfactory service. The customer in question may also have relayed the service recovery efforts through her posts on social media.

If the airline were to have analysed the post contents, it could have arrived at a better understanding of its irate flyer. That the post appeared just after midnight on a Friday could mean that the person in question must have been on a business-related travel. A 'hectic day' declaration by a person who is listed as a business executive could only confirm the nature of travel as being one for business. Also, Delhi to Bengaluru in India is definitely part of one the busiest business circuits. What were specific problems listed? There was a delay to deboard the airplane and receive luggage. How did the ground staff respond to the passenger? They showed no empathy and reacted with no urgency. What was the final outcome? The customer had threatened never to fly the airline again!

If digital were not around, such information would probably never have reached a business concern. Again, it isn't just that the digital marketplace allows for such relay of information, but it also downs inhibitions on the part of buyers. Customer-created content today has reached 'epic' proportions because buyers have been empowered by virtual settings while lowering inhibitions. No industry or business can afford to ignore such blowback on social platforms. Monitoring and responding to such content can be a start. A step ahead would be to provide buyers with a dedicated business- or brand-based virtual community platform to engage.

DISINHIBITION OF DIGITAL GROUPS

When Govind bought the Royal Enfield Himalayan Sleet Limited Edition, he was looking forward to serious off-road biking. Sadly, it wasn't to be. The first four months with the bike turned out to be a nightmare. Govind did what most buyers do when faced with an exasperating experience with a brand, taking it to social media. His post on FB read:

> Hi Guys. Please stop trusting the #royalenfield brand. It has been a very exhausting experience that I have had with it for the last 4 months. I bought the #Himalayan sleet limited edition on Feb last. Within 2 weeks, the bike started to boil from the petrol tank through the key hole on the lid. The explanation I got from the service station was, 'Sir, this is a common problem for #Himalayan sleet'. Guys I spent 2.33 lacs for the bike and see the explanation I am getting from the service people. They replaced the petrol lid and the key set after a week. In the same period bike got punctured twice going on normal road. How can they call it an off-roading bike if it gets punctured on normal roads? I had to change the tube the second time. On April, the petrol leak problem again started. Again they changed something and called me, and said it was fine. The petrol boiling issue again happened on the next day. They replaced the key set again within 4 days. Please note that you have to call them every day or else they will not update you with the status. Their attitude is 'it is your bike, deal with it'. No Royal Enfield brand love or anything. It is like some shit that you bought. Even if you complain on the #RE toll free number it is the same experience, except that they talk in a polite manner. Next again on May, the bike got

punctured again, I had to change the tube again. Please note that it took 2 days for them to change the tube. Wow, you should love these people and their f*@*@** brand value. Next day after I received the bike, I noticed petrol had started to boil out from the petrol tank, and now it is at the service centre. They have informed the factory people, so it will take them time. Please note that they won't call you to update the status, you have to call them because you know once you buy it is your crap, not 'ours'. The amount of mental agony caused by these people is beyond explainable. So guys, even if you are planning to buy a Royal Enfield go for a bullet, not this 'piece of junk'.

Once Govind's post hit the social platform, it received a substantial number of replies within no time. The comment posts came from Govind's friends circle. Most were outraged and offered suggestions on how to deal with the problem being faced. Some had their own stories to tell about their experience with the brand. When I checked on how the brand was doing with similar issues that buyers were facing, I found the problem to be a common one. In fact, just over a year ago, there was a Change.org petition titled 'Suffering Major Problems with Royal Enfield Himalayan' that had been put up. The petition got close to a thousand supporters. However, the bright part of the story is that the brand engaged with the aggrieved buyers to solve the issues they were facing, resulting in a final post that formed part of the thread. It was titled 'And They Listened', and here's what it said: 'Congratulations, Royal Enfield personnel listened to us. They are ready to rectify all the problems you have with your Himalayans. There was a shortage of spares in the service centre and it is still there but they have promised they will look into it….' What digital does to get people together, the physical can't, because of the inherent limitations the latter suffers from. At the heart of digital get-togethers for a common cause is the loosening of inhibitions at a scaled-up mass level. Individual disinhibition has a perfect foil in digital even at the aggregated level. It's not just easier to 'group up' on digital mediums; it's also the perfect foil for groups to take their message to their intended recipients in a more potent manner. The potential for virtual group formation is infinite. The people who participate can come from anywhere around the world, join any groups they want and fight for any causes they feel for. The causes aren't limited to the social sphere; they permeate the buyer–seller marketplace too. The rallying cry towards a cause can come from a single individual. Within no time, a virtual movement can be set into motion. This should of course worry business and brands. In the digital marketplace, in the business–consumer context, brands have a tendency to believe that buyers can be managed in any adverse situation because they are scattered and are far apart. The digital marketplace has undone this scenario. No buyer is isolated in the buyer–business battle when it comes to a digital back and forth. Spreading the word and having other netizens join in can happen literally within minutes and hours.

What is of course worrying is the propensity of hate groups to get together and use digital platforms to spread their ire while targeting those that take stands against such hate. The prospect of shared disinhibition that results in toxic virtual public broadcasts made by groups has the power to even pull in participants with repressed sentiments. A case in point is the response to honour killings in certain societies. Even though there are those who virtually group together to condemn such barbaric practices, there are others who do go on a toxic offensive operating in packs. These packs consist of people who see inter-caste marriages as a betrayal of the respective communities that the couple belong to. Silent until now with repressed sentiments that bristle against inter-caste alliances, digital platforms become the perfect arena to band together and go on an offensive. This isn't 'trolling'; it is a digital movement that has come to be, aided and abetted by digital technologies. Toxic group offensives that thrive in a virtual environment seep into the digital marketplace too. In 2013, when British Gas announced a 9.2 per cent rise in prices, it faced a barrage of criticism from customers. Things went from bad to worse when the company arranged for an online question and answer session on Twitter.[10] Bert Pijls, the energy company's customer services director, had logged on to take questions for an hour from customers. What followed was a barrage of critical comments from unhappy customers of the company that ranged from being funny to downright abusive. The otherwise scattered customer base of company had grouped together for the Twitter engagement and fight back against a decision they believed was unfair.

RISE OF DIGITAL COMMUNITY ENGAGEMENT

A study[11] by *Digitalist Magazine* and Oakhouse Insights of the top 50 global brands from last year provides insights into how business and brands can engage with their target buyers while making them part of a digital community. Such virtual spaces created exclusively for buyer–brand engagements ensure that there is flow of information both ways. Vanessa DiMauro, CEO of digital strategy consultancy Leader Networks, who is also an expert on online communities, says that such communities present business firms that are going digital with an opportunity to create enhanced value for their buyers. She believes that the back and forth in these communities between buyers and business enterprises can result in the generation of new ideas, solving of problems and even

---

[10] Matt West, '#PRfail: Twitter Users Give British Gas a Roasting as It Hosts Online Q&A on Social Media Site on Same Day It Hikes Bills', This Is Money, 17 October 2013. Available at: http://www.thisismoney.co.uk/money/bills/article-2465053/PRfail-Twitter-users-British-Gas-roasting-hosts-online-Q-amp-A-social-media-site-day-hikes-prices-9-2.html (accessed on 30 July 2018).

[11] Michael S. Goldberg and Christopher Koch, 'How Top Brands Nurture Their Online Communities', *Digitalist Magazine*, 16 January 2017. Available at: http://www.digitalistmag.com/customer-experience/2017/01/16/how-top-brands-nurture-their-online-communities-04835959 (accessed on 30 July 2018).

feedback on product road maps. The study found that a few companies that connected their digital community to their sales and marketing systems improved on how they tracked customer interests and responded to their needs. The 26 brands that were assessed as part of the study were rated based on four criteria, namely the way the communities engaged with members, member satisfaction, technology features and the way members engaged with the brands. The study results found that the strongest digital brand communities were those that were large in number of customers participating in and leading the discussions. Furthermore, the study found that 'community satisfaction scores were highest among companies that employed experts as moderators to answer questions, offer advice, and at times guide discussions'. Communities that did well were those where it was easy to ask questions, post comments and have fun participating. When brand engagement scores were tabulated, it was found that communities where customers were satisfied with the host company as a whole, meaning with its products and services, and the online moderators who engaged with them, scored the highest.

The *Digitalist Magazine* and Oakhouse Insights study is a proof that online disinhibition is an opportunity to engage with a community of customers. Such a digital engagement is furthered by the prospect of greater levels of participation brought about by the lowering of inhibitions. The possible scale and scope for such engagements is infinite. Imagine building global digital communities of customers. Imagine the same for other stakeholders including suppliers, channel partners and even the community at large! The potential is infinite. Companies like Social Soup in Australia have taken to building communities around product and brand usage and leveraging their potential as influencers. Social Soup currently has on board 184,914 Soupers, and in total all communities on the site put together have generated 241,772 reviews, done 838,698 trials and uploaded 204,568 photos. It's worthwhile to note that such virtual influencer communities can also be built by brand owners on their own, and the engagement generated within it can be used as platforms of influence. Who better to talk about a brand and its benefits to potential buyers than those who happily participate in and contribute to a community with the brand as its pivot! As mentioned earlier, the potential that flows from disinhibited online buyers in a digital marketplace is truly immense.

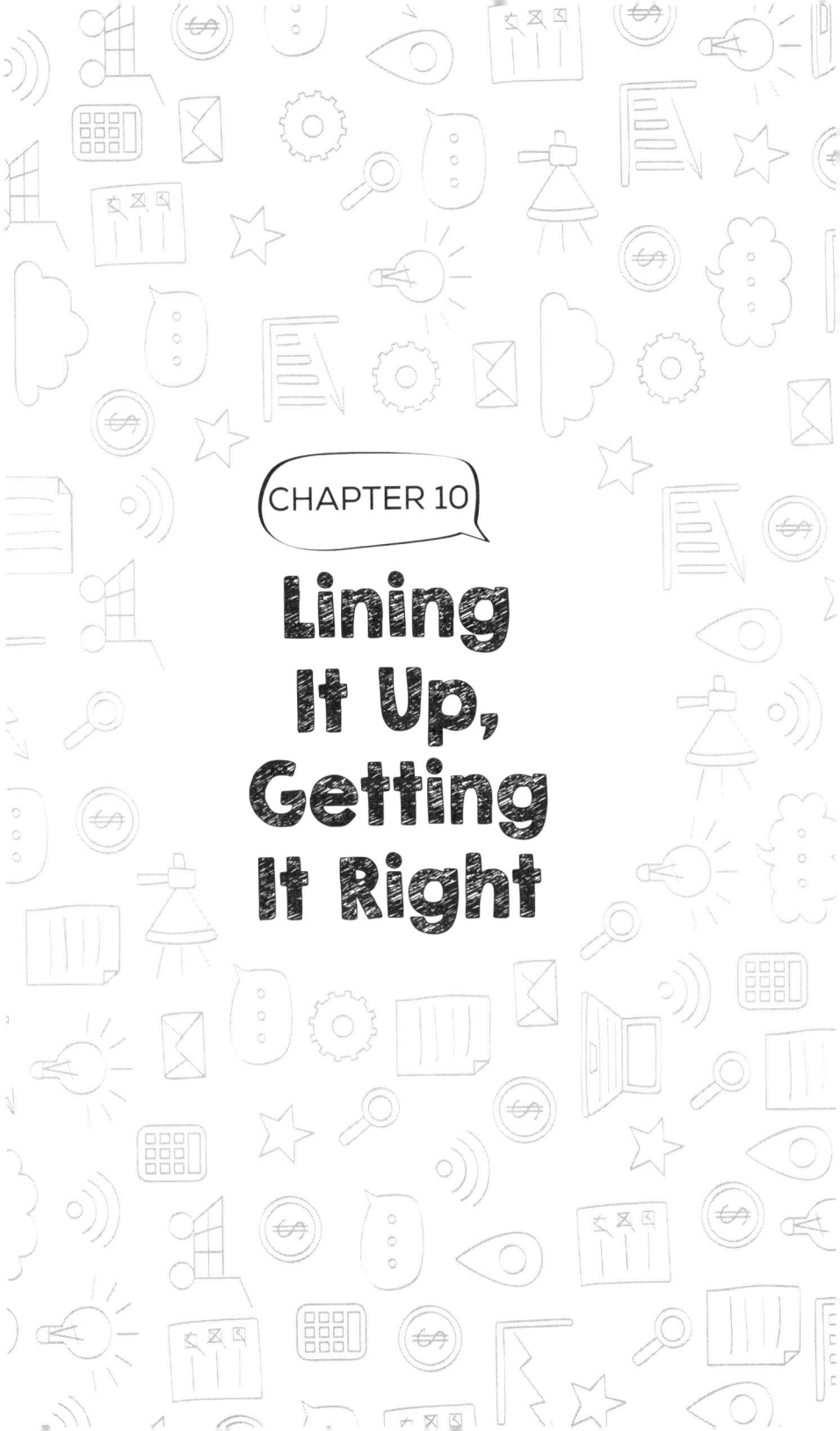

CHAPTER 10

Lining It Up, Getting It Right

When the *Houston Business Journal* announced its 2018 rankings[1] of Houston's healthiest employers, the company that made it to the top in the 'Extra large, 5,000 or more employees' category was BP America. That should not have come as a surprise to anyone, as BP America was the first company off the blocks to introduce Fitbit bracelets to its employees in the year 2013. Within two years, at least 24,500 of BP's employees were using health-tracker bracelets. At BP America, the wellness programme in place rewards employees with gift cards and other prizes for staying fit. Fitbit, the leading brand that's into products such as activity trackers and wireless-enabled wearable technology devices, has had an impressive long-term relationship with BP America. In fact, employers like BP America are now providing their workforce with activity trackers and other digital health devices to manage and maintain their health and wellness levels. It's easy to see why companies are doing this. A Duke University study[2] of 2010 published in the *Journal of Occupational and Environmental Medicine* revealed that the cost of obesity among US full-time employees was estimated to be around US$73.1 billion. The study by the university was the first one to point to a phenomenon termed 'presenteeism', which refers to lost job productivity as a result of health problems. The research study focused on three factors, namely employee medical expenditures, lost productivity on the job due to health problems (presenteeism) and absence from work (absenteeism),

[1] Jen Para, 'HBJ Announces Rankings of Houston's Healthiest Employers', *Houston Business Journal* (17 May 2018). Available at: https://www.bizjournals.com/houston/news/2018/05/17/hbj-announces-rankings-of-houston-s-healthiest.html?ana=yahoo &yptr=yahoo (accessed on 30 July 2018).

[2] Duke Global Health Institute, 'Obese Workers Cost Workplace More Than Medical Expenses, Absenteeism', *Journal of Occupational and Environmental Medicine* (7 October 2010). Available at: https://globalhealth.duke.edu/media/news/obese-workers-cost-workplace-more-medical-expenses-absenteeism (accessed on 30 July 2018).

to quantify the per capita cost of obesity among full-time workers. Corporate America realized that keeping their employees healthy both reduced costs and had a positive impact on labour productivity. This realization is why today Fitbit has been able to build up its enterprise customer portfolio that now includes big names such as Bank of America, IBM, Kimberly-Clark, Time Warner, Target and Barclays. Now it's not as if employees who double up as buyers haven't on their own bought into the fitness movement. They have, and in big numbers. The proof of buyer enthusiasm is illustrated well by the quantum of devices Fitbit shipped in 2015. That number stood at 21 million. ABI Research's prediction is that by the year 2018, more than 13 million trackers will find their way into corporate wellness plans.

The triangular relationship among Fitbit, employers and employees is an illustration of how a digital ecosystem built around health and wellness is a win-win for all of its three participants. Employers want healthier employees; employees want better health; and Fitbit with its wearables wants to be the enabler to this happening for both parties. The real enablers of course are digital devices, platforms and data. They are at the heart of what is now a health movement. There have been red flags raised on the health data capture. Such worries of digital trespassing into citizens' lives is with merit; however, the gains achieved through such digital enablement are worthy enough to allow for such intrusions with adequate checks and balances.

When Mike Stengel was serving as the general manager at the Marriot Marquis at Times Square just over a decade ago, he was beset by complaints[3] from guests who were fed up with waits for the elevator. When Chairman at the Marriot Hotels Bill Marriot suggested putting TVs in elevator landings so people could have

---

[3] Melanie D. G. Kaplan, 'Intelligent Elevators Answer Vertical Challenges', ZDNet, 17 July 2012. Available at: https://www.zdnet.com/article/intelligent-elevators-answer-vertical-challenges/ (accessed on 30 July 2018).

something to watch as they waited, Mike disagreed with the idea. He pointed out that if the hotel did that, people would then know how much they had waited because they would have watched an entire show during the wait! The elevator problem was a big one at the Marquis because the 49-storey hotel, which had almost 2,000 guest rooms, was designed and built largely for meetings. To fix the problem that was causing the Marquis losses in the business of meeting and lowering satisfaction scores, Mike turned to Schindler Elevator Corp., the North American operation of the Swiss-based Schindler Group. The advice he got from the elevator company was to install the 'Miconic 10' at the Marquis, a smart elevator system that would allow users to enter their floor onto a keypad. The app would calculate the fastest time to destination and accordingly assign an elevator. The promise to Mike was that the smart elevator would increase efficiency by 15–20 per cent. When the smart elevator project was completed and put into effect, the results were remarkable. The efficiency of the elevator system went by a whopping 50 per cent! The complaints Mike received after the new elevator system went live? None! The Miconic 10 and other such smart elevators around the world are transforming the way we ride up and down buildings with multiple floors. When such technology is deployed in places like hospitals, the time saved and the pace of movement achieved can even literally save lives. The elevator wait problem is not a simple one. In fact, its consequences are grave in terms of productivity lost. A 2010 IBM Smarter Buildings study[4] revealed that the cumulative time that office workers spent stuck in elevators in a year totalled 33 years across the 16 cities in the United States. The city of New York topped with a loss of 5.9 years. The other cities panned out thus: Los Angeles, 4.3 years; Chicago, 3.2 years; Houston, 2.9 years; Dallas/Fort Worth, 2.4 years; Washington, DC, 2.2 years; Atlanta, 1.9 years; Boston, 1.8 years; Philadelphia, 1.7 years; San Francisco/Oakland/San Jose, 1.4 years; Detroit, 1.1 years; Seattle/Tacoma,

---

4 IBM, 'IBM Survey Shows Strengths, Gaps in U.S. Office Buildings', IBM, 29 April 2010. Available at: https://www-03.ibm.com/press/us/en/pressrelease/30191.wss (accessed on 30 July 2018).

1 year; Denver, 1 year; Phoenix/Prescott, 0.8 year; Tampa/St Petersburg, 0.6 year; and Minneapolis/St Paul, 0.5 year.

Intelligent elevators and systems have done much to eliminate such loss of productivity and brought it down to negligible amounts of time lost. Of course, it isn't just smart elevators that know us and about where we are headed as we ride them, but it's also smart rooms, homes, buildings and even cities. Every digital footprint we leave behind us via connected devices and platforms allows for a digital-enabled system to know us better, and therefore customize, for example, services based on who we are and the way we behave. This in effect raises the issue of a trade-off between privacy and enhanced convenience and efficiency, and that is where 'digital alignment' steps in as the core pursuit. That industries are using digital technologies to enhance their performance is a taken. What digital technologies can bring about is best summed in Accenture Institute for High Performance's 2014 report titled *From Looking Digital to Being Digital*. In a note on 'real-time adaptation', the report states[5],

> Pervasive digital connections between systems, people, places and things—sometimes referred to as the 'internet of everything'—will produce a dynamic flow of digital information about where machines and people are, what they are doing, and how they are doing. Intelligent assistants will use this information to help employees make smart decisions even when they cannot calculate the implications of all that data themselves. The potential for dynamic and speedier decision-making will bring greater levels of operational flexibility and productivity to industries.

The application of digital technologies is without limitations across business verticals. In many ways, its potential to transform is almost infinite. Industries and business may currently be at

---

[5] https://www.peoplematters.in/article/hrs-digital-transformation/role-of-hr-in-digital-transformation-at-accenture-13849?utm_source=peoplematters&utm_medium=interstitial&utm_campaign=learnings-of-the-day (accessed on 11 September 2018).

different stages in this transformation process. In the health care industry, for example, doctors are increasingly embracing the power of digital to provide better care to their patients. In the 2015 survey by Indigene,[6] conducted on a sample of 1,600 plus doctors across the United States, India, China, Japan and the rest of Asia to uncover their 'digital preferences', it was found that 60 per cent doctors preferred tablet-based detailing (use of an electronic detail aid on a Tablet PC by sales reps in a face-to-face discussion with doctors), while the rest did the same online (providing pharmaceutical sales presentations to physicians remotely via digital channels). Of the doctors surveyed, 34 per cent preferred using smartphones, while 26 per cent went with the use of laptops. Over 64 per cent of doctor–patient communication was conducted via digital engagements across all markets. The top 3 digital channels used for such communications were email, via the website and text messages. Writing in *Forbes*, Daniel Newman listed the current digital transformation trends in health care as telemedicine, mobility and cloud access, wearables and IoT, and artificial intelligence and big data, all of which ensure that there are more empowered customers aka patients. All of the digital advances are ensuring that the business of health care is now operating in an infinite marketplace with problems of accessibility being breached. Telemedicine, for example, is allowing for caregivers to reach patients in the remotest of corners. The wearables market is ensuring people to keep an eye on their health indicators and transmit the numbers when required in an instant to a physician anywhere around the world. The combination of a business looking out for employees so they remain fit and productive and a business looking to better customer solutions is the key to getting business–buyer alignment right.

Digital alignment is a subset and an enhancer to business–buyer alignment. Digital technology and its applications enable better decision-making both within and on the outside for business firms.

---

[6] Gaurav Kapoor, 'Doctors and Digital Devices: India Bridging the Gap', NASSCOM, 16 February 2016. Available at: http://blogs.nasscom.in/doctors-and-digital-devices-india-bridging-the-gap/ (accessed on 30 July 2018).

The number of firms around the world that are making a dash towards digital embracement is on the upswing (see Table 10.1). The IDG 2018 'State of Digital Business Transformation' study[7] suggests that start-up businesses have been quicker on adopting digital than those that have been around for a while. Almost all of the start-ups that are up and running (95%) have digital plans in pace as against 87 per cent of traditional enterprises founded 50 years ago or later. A little over one in two start-ups have adopted a digital business strategy in comparison to 38 per cent of established firms.[8]

If business and marketing are about bettering people's lives as employees and customers, then digital is the perfect ally to getting there. An aligned business is one that is aptly designed to get the best of their employees so they can in turn build and deliver superior value propositions to their customers. This understanding is the key to getting digital adoption right. Michael Krigsman, a digital transformation expert and host of CXOTalk, points to the fact that most companies wanting to use the power of digital to transform their business often start with targeted areas like the IT function. He says that very few companies are willing to 'completely upend what they are doing'. He recommends that at some point, businesses must move from mere operational improvements or as he terms it 'departmental transformation' to capturing wholly the immense opportunities that digital technologies are making possible. In one of Michael's CXO talks,[9] he engaged with Chris Satchell, executive vice president and chief product officer at

---

[7] IDG, 'Understand How Organizations Evolve to a Digital Business Model' (Digital Business Survey, 2018). Available at: http://resources.idg.com/download/white-paper/2018-digital-business (accessed on 30 July 2018).

[8] Louis Columbus, 'The State of Digital Business Transformation, 2018', Forbes, 22 April 2018. Available at: https://www.forbes.com/sites/louiscolumbus/2018/04/22/the-state-of-digital-business-transformation-2018/#5bd0d4c05883 (accessed on 30 July 2018).

[9] CXOTalk, 'Comcast: Digital Transformation and Innovation', CXOTalk, 2018. Available at: https://www.cxotalk.com/episode/comcast-digital-transformation-innovation (accessed on 30 July 2018).

Table 10.1 Digital Strategies and Technologies in Use/to Be Used

Tools	On My Radar/ Actively Researching (%)	Piloting New Initiatives (%)	In Production in a Business Unit or Division (%)	In Production Enterprise-wide (%)	Upgrading/ Refining (%)	Not Interested (%)
Artificial Intelligence	39	17	11	5	3	27
Machine Learning	34	21	9	5	4	27
IoT	31	19	13	9	5	22
Software-defined Networking	30	15	14	11	4	26
AR/VR	29	14	7	3	2	44
Software-defined Storage	28	16	14	8	4	29
App Performance Monitoring Technology	27	16	18	14	5	20
Bots	26	13	11	5	1	44
Microservices/Containers	25	15	15	7	4	33
APIs/Embeddable	22	15	19	17	4	23
Private Cloud	20	15	20	26	7	12
Public Cloud	20	15	20	21	4	19
Mobile Technology	18	18	22	27	10	6
Big Data/Analytics	15	20	26	24	9	6

Source: IDG, 'Understand How Organizations Evolve to a Digital Business Model'.

Comcast Cable, to explore more on the digital transformation that was changing the cable TV industry. When Chris was asked about digital transformation at Comcast, and its connected human and cultural dimensions, here's what he had to say:

> Culture and people are the keys of any technology transformation, I mean any transformation, but especially technology transformations. One of the things I always think about, and I talk to people about this, is in my whole career I've only actually ever seen two projects go off the rails because of sheer technology, and your technology is just too hard to overcome. I mean it's just limitations at that point in history where you just couldn't get past it. Every other issue has been about people, and so, for me, technology is inherently a people issue. That's how we approach it. You have to get the culture right. You have to get the context right. You have to get the team right. You have to get direction right. Then you can drive change and accelerate it.

In an IDG study[10] from a couple of years ago on digital transformation, it was found that a little over 80 per cent of the executives surveyed pointed to enhance operational efficiency as the most important driver of the digital transformation efforts. What followed were business agility, higher employee productivity and stronger competitive advantage. Of the focus areas of transformation, a little over one in two of the surveyed executives admitted to IT being one of the first functions to be targeted. The road beyond IT has been a slow and arduous one for most companies. Of the 40 per cent of surveyed companies that have tried to move beyond IT to the operations function, only 27 per cent have been able to make some significant headway, and 25 per cent revealed

---

[10] IDG and Dell, 'Digital Transformation: Crossing the Chasm from IT to the Business' (white paper, 2017). Available at: http://marketing.dell.com/Global/FileLib/eLearning/Digital-Transformation-Crossing-the-Chasm.pdf (accessed on 30 July 2018).

that it was one of the most difficult areas to transform. An almost similar number felt that the finance function was a difficult area and only 14 per cent of those surveyed reported to have made some progress. The reasons for the lack of progress with adopting digital ranged from lack of time and resources, and the difficulty in acquiring the required skills and expertise. Two-thirds of those surveyed admitted to the roadblock of 'mindsets'. It was difficult to change existing cultures and mindsets, and what added to the problem was a clash between millennials who were keen to push digital adoption and the older ones who were cautious of technology from a cultural standpoint and therefore were unwilling to let go of the material they had created decades ago.

The mindset to adopt digital technologies in customer-facing functions has also not been encouraging. In the IDG survey, only 20 per cent of those surveyed named sales as a transformation focus area. An even lower 13 per cent reported any significant progress with the sales function (see Table 10.2).

Table 10.2 Importance of Key Digital Transformation Drivers

Digital Drivers	Critical (%)	Very Important (%)	Somewhat Important (%)	Not Very Important (%)	Not at all Important (%)
Increasing Operational Efficiency	31	51	14	2	1
Increasing Business Agility	34	42	18	5	1
Increasing Employee Productivity	22	52	22	4	1
Gaining Competitive Advantage	35	37	14	10	4
New Revenue Growth	28	35	20	9	7

Source: IDG (166 respondents to the question, 'How important are the following as drivers of digital transformation at your organization?').

The degree of digital alignment between the core and support functions within a business to that of value creation for target customers is still at a nascent stage. For organizations that intend to get digital alignment right, the starting point and the pivot needs to be the digital buyer. To work forwards from the buyer, it's important to recognize the difference between being 'digital driven' and to 'drive digital'.

DIGITAL ALIGNMENT

The seeds of the Uber taxi aggregator idea were sown at the 2008 LeWeb conference in Paris when long-time pals Travis Kalanick and Garrett Camp were ruminating over the many painful things they had endured in life including finding and hailing a cab in the most trying of circumstances. Within two years, Travis and Garrett were testing the idea out by plying a few black cars in the Big Apple. The city that followed New York was San Francisco and soon travellers were marvelling at this amazing new way of calling for and riding in cabs. What followed for Uber and its aggregator taxi service is now part of business legend. For the record, Uber started with a Series A funding of US$11.5 million, followed by an additional US$32 million in the second round of funding, which then was backed up by a whopping US$1.2 billion in the final round of funding. Uber entered India in 2013 operating cabs in the city of Bengaluru. By 2016, it had added 27 more cities roping in 400,000 driver partners. Despite multiple controversies, the cab aggregator has only been growing in strength. As of May 2018, Uber reported that it had net revenues of US$2.5 billion, which was a 67 per cent increase year-over-year on gross sales of US$11.3 billion. Estimates put the company's valuation at US$62 billion.

Uber and its radically new way of enabling riders to hail cabs and in the process eliminate the 'trying' ways they had been using to do the same for long is an apt illustration of 'driving digital'. The firms that drive digital adoption by buyers disrupt existing models of value creation to replace it with a completely new way of solving buyer needs and problems. Being 'digital driven' on the other hand is about using the power of the digital ecosystem to better existing products and services. So when a service provider sets up an online presence and complements the existing physical format of ordering or delivery, the buyer is being provided with enhanced conveniences. Such digital provisions flow from a change in the buying behaviour of the service provider's target customers. The change in behaviour is especially due to buyers migrating to digital platforms and using digital devices and apps for various personal and professional purposes. So when Indian Railways sets up a ticketing portal online, what it does is eliminate the hassles train travellers in India have had queuing up for tickets at physical counters that are sparse in their numbers. What digital-driven companies do is follow the changes brought about by digital adoption. What firms that drive digital do is disrupt the way value is created replacing it with a radically new digital-enabled value proposition, prompting buyers to abandon existing products and services and shift to this pioneering solution. The latter set of companies change buying behaviour with their innovative use of digital technologies, whereas the former adapt to changes buyers have taken to on account of their going digital. According to David Yockelson, Research vice president at Gartner, an American research and advisory firm, 'Digital disruptors are organizations that have "taken advantage of digital capabilities in one form or another to create and drive fundamental shifts—intentionally or otherwise—in their market and perhaps others … either by design or as a secondary effect"'. The companies David calls disruptors are in effect those firms that 'drive digital' in the marketplace and get buyers to adopt the value outcomes of their disruptive use of digital competencies. According to Gartner,[11] the four elements

---

[11] Janelle B. Hill, 'Leading Through Digital Disruption', Gartner, 2017. Available at: https://www.gartner.com/imagesrv/books/digital-disruption/pdf/digital_disruption_ebook.pdf (accessed on 30 July 2018).

that make up digital disruptions are business, technology, industry and society. The business part of digital disruption refers to radical changes to market development, pricing, delivery and other such value-creating functions. Technology refers to issues such as invention, design, usage and other such factors, while the industrial element encompasses issues such as processes, standards, methods and customers. Disruptions at the social level have macro implications for culture, habits and norms, and they may even engineer sociocultural movements.[12]

One of the dilemmas that businesses face in embracing digital is whether to leverage it to better their value propositions for their target buyers or to use it as a disruptive tool to build a completely new business model. Take the case of Blockbuster, a bricks-and-mortar video rental firm, versus Netflix. The latter first stepped in as a competitor by sending out DVDs by mail and then moved to streaming video on demand. This DVD-mailing act combined with anytime returns eliminated the problems that customers were facing with renting videos from Blockbuster and having to return it by a required date. In addition, the move that Netflix made from a monthly rental fee to individual rental aided by its digitized business model had a significant impact on customer demand for its services. Within no time, Blockbuster began losing customers and had to file for bankruptcy. When the very method of renting video content for entertainment is disrupted via digital technologies, firms like Blockbuster can't do much. To fight back in a case of disruptive change may in all possibility require an abandonment of the current model of value creation. That isn't easy considering the kind of strategic commitments already made to operate the existing model. Contrast this with when the use of digital by a competitor results in an incremental upping of value for the buyer. In such a case, it's easier for incumbents to catch up and imitate. A case in point is bricks-and-mortar retailers scrambling to add digital stores to complement what they were doing offline. The way to not go wrong and get swept out of a digitized marketplace is to religiously stick to the basics of tracking and adapting to changes. It's equally important to hold on to the fundamentals of value creation. To the customer, enhanced value comes from a better solution that is either incrementally or radically better than existing ones. In strategy parlance, the way to get to that is to either lower costs or add more differentials and pass such value adds to buyers in the form of either lower prices or differentiated products and services. Any superior solution to the buyer is one that solves a need or problem in an unmatched manner when compared to other available solutions in a market-place. If you were to closely track the changes digital has wrought in any marketplace, you will see that it has either lowered costs the customer incurs or provided enhanced benefits. Now the picture of better value creation is completed only when it is understood within the context of a marketplace that consists of buyers who are trying to maximize the value they seek from the solutions they buy into. Such value sought not only operates on utilitarian lines but also has emotional and psychological elements associated with it.

To understand this best, turn your focus on to the business of online grocery. Data from different markets suggest that online grocery formats have been struggling until now. The grocery business in the United States is pegged at US$675 billion, and currently is dominated by physical stores. Online grocery accounts for hardly 1–3 per cent of industry sales. According to RedSeer Consulting, the online grocery market in India stood at US$1 billion in 2017. Despite exits by a few start-ups that tried to scale up their online grocery business in India, the numbers seem to be encouraging. Goldman Sachs predicts the market to grow to US$40 million by 2019, which would register a CAGR of 62 per cent from 2016 to 2022. Morgan Stanley estimates put the online grocery segment to be among the fastest growing markets in India, increasing at a CAGR of 141 per cent by 2020. This would in effect contribute US$15 billion or 12.5 per cent of overall online retail sales. The reasons why online grocery has not taken off well in the United States is due to the twin problems of perishability and price. The first problem is why grocery shoppers want to pick the merchandise themselves, not trusting others to do it as well as they do. The higher labour and delivery costs associated with grocery buying online cannot be passed on to buyers as they are unwilling to pay for the same. The inability to pass on such costs is why online grocery selling leaves the sellers with wafer-thin margins to work with. In the United States, Amazon is attempting to solve these problems by building a digitized physical store christened 'Amazon Go'. Now that's one way of using digital to keep customers. However, this may work in a large, organized and developed market like the United States. India on its part throws up challenges that are uniquely different from those in the West. Although the two problems mentioned earlier stayed intact for India, there are others unique to the country that online grocery solved for shoppers in India. My research on shoppers who are regular buyers of online grocery revealed some fascinating insights. The switch from physical stores happened as an outcome of combined reasons of experimenting with the digital format, and the promotions that lured them in. The sheer convenience of getting groceries home was an overwhelming one. Clogged cities with terrible infrastructure and traffic meant that these shoppers were reluctant to drive to big stores located some distance away. The parking problems and charges added to their woes. The time taken to shop and checkout again wasn't something they were looking forward to. Digital grocery buying saved them time that they could utilize for other pursuits. Added to this was the fact that over time, these shoppers have learned to shift their trust to shop personnel who they believed would pick the right stuff for them, and if they didn't, they could always ask for refunds or even return it. With Amazon operating the 'Amazon Now' (recently rebranded as Prime Now) format in India, digital grocery shoppers could pick stuff from large stores close by via the app. What these Indian shoppers also did was rely on nearby mom-and-pop stores for emergency purchase. When they bought online, they usually picked up grocery in larger quantities. Even in the United States, online grocery is now showing signs of taking off. The latest data from the Food Marketing Institute and Nielsen study reveal that nearly one in two Americans now buy groceries online. It was found that 49 per cent of grocery shoppers in the United States had purchased consumer packaged goods online. Among millennials, that number stood at 61 per cent and for Gen Xers it was 55 per cent. In terms of the population buying online, the numbers have doubled from the previous year.

Digitizing retail isn't only about enhancing utilitarian value. There are varied ways to engage and connect at an emotional level with target buyers driving value that is beyond the utilitarian. Take the case of Licious, an Indian online meat shop that reached out to its buyers on World Environment Day with an appeal. The emailer they sent out was titled 'Here's Our #LetterOfChange'. Part of the contents of the letter read thus, 'Get our #LetterOfChange seed paper with your order today. Pot it. Water it. Watch it Grow'. The online meat-selling company was encouraging people to contribute towards caring for the environment by growing a plant, the seeds (in paper form) for which were being delivered along with the meat orders. If Licious' little gesture was about doing a bit more than just sell meat online, pizza maker Domino's went all out to leverage digital platforms as engagement tools. Pizza brand Domino's 'AnyWare' initiative is another example of using digital to make the pizza-ordering activity a lot of fun. Under this initiative, anytime you feel like ordering a Domino's pizza, you can tap your Pebble or Apple smartwatch, or ask Alexa, or even tweet or text a pizza emoji. The pizza company on its part is ensuring its doing everything digital possible to engage with the brand and of course order pizzas. As of now, pizza lovers can place orders using any of the following AnyWare platforms: Google Home, Amazon

Alexa, Slack, Messenger, zero-click app, text, tweet, Ford SYNC, Samsung Smart TV, voice ordering with DOM and smartwatches. When Kelly Garcia, Domino's SVP of eCommerce Development and Emerging Technologies, was asked how the pizza company got to using digital in such innovative manners, he pointed to the need for a culture of collaboration and alignment between the IT and marketing teams. Kelly revealed that the digital transformation the company tasted was an outcome of two distinctive phases the initiative went through. The first step was to get the fundamentals right. Adopting a mobile-first approach, the company built top-notch native and iOS mobile clients coupled with a superb responsive experience. The move to mobile was natural as digital sales were ramping up with mobile's contribution also going on the upswing. The first phase was about building a strong foundation through mobile excellence and testing done to figure what worked. Customer profiles were also built to ensure that targeting was done right. During the second phase, Domino's focused on surprising and delighting their customers. The innovative AnyWare ordering platform was built so that pizza lovers could order on their favourite digital devices any way they wanted and anywhere. A loyalty programme was also introduced to keep customers from switching.[13]

One part of digital alignment is about leveraging the power of digital in an integrated manner so that business functions can generate enhanced value. The front-facing part of the alignment is using digital platforms, tools and technologies to connect and engage with customers so as to build deeper long-term relationships and as outcome of that enjoy accruals of lifetime value.

---

[13] https://www.forbes.com/sites/kylewong/2018/01/26/how-dominos-transformed-into-an-ecommerce-powerhouse-whose-product-is-pizza/#218f6ec47f76 (accessed on 11 September 2018).

Epilogue

Speaking to Fox Business Network's Maria Bartiromo, here's what President Donald Trump said about the power of social media[1]:

> I doubt I would be here if it weren't for social media, to be honest with you. I have friends that say, Oh, don't use social media. See, I don't call it tweets. Tweeting is like a typewriter—when I put it out, you put it immediately on your show. I mean, the other day, I put something out, two seconds later I am watching your show, it's up.

Trump was right. Without a platform that connected him directly with his base of voters, he would not have sailed into the White House as the 45th president of the United States.

To truly understand the power that the digital medium holds, you only have to look at the man in the White House. If you still aren't convinced, you should check on the man behind the Bieber fever, Justin. It's possible that you may not agree on YouTube turning Justin into the money-making music machine that he is. You may even point to Usher and Island Def Jam Recordings as those who built Justin's recording career, but here's what you are missing. If 12-year-old Justin did not have those videos out on YouTube that got in subscribers by the thousands, no one would have noticed the boy,

[1] https://www.independent.co.uk/news/world/americas/us-politics/donald-trump-tweets-twitter-social-media-facebook-instagram-fox-business-network-would-not-be-a8013491.html (accessed on 11 September 2018).

leave alone invest in his music and money-making potential. Now consider what Donald was up against in the US presidential elections. With a fraction of resources vis-à-vis his Democrat opponent, it wasn't going to be easy getting what he stood for across to his voters. Sure, he was a celebrity in his own right, but appealing to voters is nothing like being popular via a TV show. But win he did, against mainstream media opposition and polling predictions. This is what Carrie Levine, Michael Beckel and Dave Levinthal wrote in a 2016 *Time* magazine article after the Presidential elections[2]:

> In the end, Donald Trump defeated big money. The Republican's presidential campaign raised less than half of what Democratic nominee Hillary Clinton did. He ran a fraction of the TV ads, even in decisive battleground states. And although prominent Republican donors came to Trump's aid during the campaign's final days, his supportive super PACs and other political groups raised relatively paltry sums when compared to Clinton's groups. Somehow, for the volatile, unpredictable New York businessman, that was enough on his way to winning the White House. More than enough. A swell of white, working class voters, who believed the billionaire when he said he wouldn't be beholden to special interests, propelled him in a manner that defied polls and the predictions of pundits relying on conventional wisdom for how to win elections.

If the authors of the *Time* article had dug deeper, they would have found that what carried Trump across the finish line and what was enough was the connect the man had cemented between him and his base of voters. He achieved this connect by directly speaking to them, and by bypassing mainstream media outlets except

---

2 http://time.com/4563949/donald-trump-hillary-clinton-money-machine-election/ (accessed on 11 September 2018).

for Fox. What aided Trump were digital and social platforms. The digital sites that carried the Trump message across were Drudge Report, Breitbart, the Gateway Pundit and many others. Here's an example of the number of people Trump connects on his own via digital media. His followers on Twitter now stand at 52.9 million. According to the Alliance for Audited Media figures, Donald Trump's Twitter followers far exceed the number of subscribers for all news outlets, print and digital, in the United States—35 million on weekdays and 38 million on Sundays as of 2016.

The Trump and Bieber phenomena must open our eyes to seeing digital as truly a tool for empowerment. When Thomas Friedman wrote about the world being flat, he was talking about globalization. If globalization opened up border gates for resource, material, technology and knowledge flows, digital requires none of that to turn the globe flat. In an infinite digital world, borders and gatekeepers don't matter. Big media and money count for nothing as in the case of Donald. The digital world is truly flat in that it allows for all-round and inclusive participation without hindrances of time and geography. In terms of empowerment, digital is the triumph of people over corporations. It is the coming of age for the buyer who can now seek out what is the best value out there in a competitive marketplace. Digital is also about new types of corporations, sellers and marketers, ones who can connect to their audiences in transparent ways and engage in manners that are truly democratic.

In the end, digital is about people and the power of technology in making their lives better. Those who succeed in the new world of digital will be those who never take their eyes off what this phenomenon truly represents, a way to make the world a better place for all its inhabitants.

Bibliography

A. T. Kearney, and WHU. 'Digital Supply Chains: Increasingly Critical for Competitive Edge'. European A. T. Kearney/WHU Logistics Study 2015. Available at: https://www.atkearney.com/documents/20152/435077/Digital%2BSupply%2BChains.pdf/82bf637e-bfa9-5922-ce03-866b7b17a492 (accessed on 17 July 2018).

American Press Institute. '"Who Shared It?" How Americans Decide What News to Trust on Social Media'. American Press Institute, 20 March 2017. Available at: https://www.americanpressinstitute.org/publications/reports/survey-research/trust-social-media/ (accessed on 17 July 2018).

Angel, Gary. 'Change or Die: Lessons from the Retail Apocalypse'. Digital Mortar Blog, 4 May 2017. Available at: http://digitalmortar.com/change-die-lessons-retail-apocalypse/ (accessed on 17 July 2018).

Asian News International. 'India Takes the Organic Path!' Republic, 6 April 2018. Available at: https://www.republicworld.com/lifestyle/fashion/india-takes-the-organic-path (accessed on 18 July 2018).

B2X Care Solutions GmbH. 'Smartphone Obsession Grows with 25% of Millennials Spending More Than 5 Hours per Day on the Phone'. B2X, 18 May 2017. Available at: https://globenewswire.com/news-release/2017/05/18/987607/0/en/Smartphone-Obsession-Grows-with-25-of-Millennials-Spending-More-Than-5-Hours-Per-Day-on-the-Phone.html (accessed on 26 July 2018).

Blodget, Henry. 'CEO of "China's Apple" Is Insulted by Comparison to Apple—Says They're More Like Google or Amazon'. Business Insider, 15 August 2013. Available at: http://www.businessinsider.com/ceo-of-chinas-apple-xiaomi-lei-jun-2013-8?IR=T& (accessed on 17 July 2018).

Braiker, Brian. 'The "Ice Bucket Challenge": A Case Study in Viral Marketing Gold'. Digiday, 14 August 2014. Available at: https://digiday.com/marketing/ice-bucket-challenge-case-study-viral-marketing-success/ (accessed on 25 July 2018).

Breitbart News. '#DumpKelloggs: Breakfast Brand Blacklists Breitbart, Declares Hate for 45,000,000 Readers'. Breitbart, 30 November 2016. Available at: http://www.breitbart.com/big-government/2016/11/30/dumpkelloggs-kelloggs-declares-hate-45-million-americans-blacklisting-breitbart/ (accessed on 18 July 2018).

Brennen, Sofia. 'The (VERY Wealthy) Bank of Mum and Dad: Rich Parents of Instagram Put Their Children in the Shade with Their Boastful Photos of Pet Cheetahs, Private Jets and Wads of Cash'. *The Daily Mail*, 7 June 2016. Available at: http://www.dailymail.co.uk/femail/article-3629123/Rich-parents-Instagram-lavish-lifestyles.html (accessed on 27 July 2018).

———. 'Are These the Most Outrageous Displays of Teenage Wealth Yet? Luxury Kids of Instagram Flaunt Their VERY Lavish Lifestyles—Including Pouring Champagne over Cereal and a Pet Tiger'. *The Daily Mail*, 4 October 2017. Available at: http://www.dailymail.co.uk/femail/article-4944804/Luxury-Kids-Instagram-flaunt-lavish-lifestyles.html (accessed on 27 July 2018).

Campaign India Team. 'The Rise and Rise of the Influencer'. Campaign India, 17 November 2017. Available at: http://www.campaignindia.in/article/the-rise-and-rise-of-the-influencer/441102 (accessed on 19 July 2018).

Cheng, Shirley Y. Y., Tiffany Barnett White, and Lan Nguyen Chaplin. 'The Effects of Self-brand Connections on Responses to Brand Failure: A New Look at the Consumer–Brand Relationship'. *Journal of Consumer Psychology* (15 June 2011). Available at: https://onlinelibrary.wiley.com/doi/pdf/10.1016/j.jcps.2011.05.005 (accessed on 27 July 2018).

Columbus, Louis. 'The State of Digital Business Transformation, 2018'. *Forbes*, 22 April 2018. Available at: https://www.forbes.com/sites/louiscolumbus/2018/04/22/the-state-of-digital-business-transformation-2018/#5bd0d4c05883 (accessed on 30 July 2018).

Common Sense Media. 'The Common Sense Census: Media Use by Tweens and Teens'. Common Sense Media, 2015. Available at: https://www.commonsense media.org/sites/default/files/uploads/research/census_researchreport.pdf (accessed on 19 July 2018).

Court, David, Dave Elzinga, Susan Mulder, and Ole Jørgen Vetvik. 'The Consumer Decision Journey'. McKinsey, June 2009. Available at: https://www.mckinsey.com/business-functions/marketing-and-sales/our-insights/the-consumer-decision-journey (accessed on 26 July 2018).

Cranfordteague, Jason. 'Digiphrenia—Excerpt from Douglas Rushkoff's Present Shock'. *Wired*, 26 March 2013. Available at: https://www.wired.com/2013/03/digiphrenia-excerpt-from-douglas-rushkoffs-present-shock/ (accessed on 19 July 2018).

CXOTalk. 'Comcast: Digital Transformation and Innovation'. CXOTalk, 2018. Available at: https://www.cxotalk.com/episode/comcast-digital-transformation-innovation (accessed on 30 July 2018).

D'Arnault, Clayton. 'What Is Digital Culture?' *Digital Culturist*, 20 October 2015. Available at: https://digitalculturist.com/what-is-digital-culture-5cbe91bfad1b (accessed on 27 July 2018).

Deloitte, and Retailers Association of India. 'Trend-setting Millennials: Redefining the Consumer Story'. Deloitte, February 2018. Available at: https://www2.deloitte.com/content/dam/Deloitte/in/Documents/CIP/in-cip-trendsetting-millenials-noexp.pdf

Doctoroff, Tom. 'The False Divide Between Digital vs. Traditional Media'. *Huffington Post*, 12 May 2015. Available at: http://www.huffingtonpost.com/tom-doctoroff/the-false-divide-between_b_8730234.html (accessed on 25 July 2018).

Dodez, Aaron. 'User-generated Video Content Is Drowning Out Brands: But There Is Hope'. Tubular Insights, 7 May 2015. Available at: http://tubularinsights.com/user-generated-content-brand-videos/ (accessed on 25 July 2018).

Dredge, Stuart. 'How Toy Unboxing Channels Became YouTube's Real Stars'. *The Guardian*, 28 April 2016. Available at: https://www.theguardian.com/technology/2016/apr/28/children-toys-unboxing-channels-youtube-real-stars (accessed on 27 July 2018).

Duke Global Health Institute. 'Obese Workers Cost Workplace More Than Medical Expenses, Absenteeism'. *Journal of Occupational and Environmental Medicine* (7 October 2010). Available at: https://globalhealth.duke.edu/media/news/obese-workers-cost-workplace-more-medical-expenses-absenteeism (accessed on 30 July 2018).

Ehrenberg, Rachel. 'Facebook Peer Pressure Gets Out the Vote'. Science News, 12 September 2012. Available at: https://www.sciencenews.org/article/facebook-peer-pressure-gets-out-vote (accessed on 19 July 2018).

Elder, Robert. 'More Young People Are Watching Less Traditional TV'. Business Insider, 12 July 2016. Available at: http://www.businessinsider.com/more-young-people-are-watching-less-traditional-tv-2016-7?IR=T (accessed on 18 July 2018).

eMarketer. 'Adults Spend Half of Daily Media Usage on Digital'. Inside Radio, 10 October 2017. Available at: http://www.insideradio.com/free/emarketer-adults-spend-half-of-daily-media-usage-on-digital/article_24addc46-ad97-11e7-bbda-4f926ba26027.html (accessed on 27 July 2018).

Englander, Elizabeth. 'Cyberbullying & Bullying in Massachusetts: Frequency & Motivations'. Research Brief, Massachusetts Aggression Reduction Center, MARC Publications, 2008. Available at: http://vc.bridgew.edu/cgi/viewcontent.cgi?article=1009&context=marc_pubs (accessed on 19 July 2018).

Ersdal, Hannah, and Sølvi Svendby. 'Privacy and Social Media: Do Users Really Care?' Norwegian University of Science and Technology, 2016. Available at: https://brage.bibsys.no/xmlui/bitstream/handle/11250/2403229/15179_FULLTEXT.pdf?sequence=1 (accessed on 30 July 2018).

Esseveld, Nelson. 'Should You Co-create Your Content with Influencers?' TapInfluence, 22 March 2017. Available at: https://www.tapinfluence.com/co-create-content-influencers/ (accessed on 19 July 2018).

Eugene Tang, Chua Kong Ho, and Yingzhi Yang. 'Rise of Xiaomi: The Chinese Start-up Poised to Become World's Biggest IPO of 2018'. *South China Morning Post*, 18 April 2018. Available at: http://www.scmp.com/business/companies/article/2142130/rise-xiaomi-inside-humble-chinese-start-poised-become-years (accessed on 26 July 2018).

FE Online. 'Gone in Seconds! Xiaomi Says Sold Over 300,000 Redmi Note 5, Note 5 Pro Units Within 3 Minutes'. *Financial Express*, 23 February 2018. Available at: https://www.financialexpress.com/industry/technology/gone-in-seconds-xiaomi-says-sold-over-300000-redmi-note-5-note-5-pro-units-within-3-minutes/1076785/ (accessed on 26 July 2018).

Fullerton, Kevin. 'Hashtag Humiliation: The Tweets That Shamed a Brand'. Attercopia, 19 April 2016. Available at: https://www.attercopia.co.uk/2016/04/19/hashtag-humiliation-the-tweets-that-shamed-a-brand/ (accessed on 30 July 2018).

Goldberg, Michael S., and Christopher Koch. 'How Top Brands Nurture Their Online Communities'. *Digitalist Magazine*, 16 January 2017. Available at: http://www.digitalistmag.com/customer-experience/2017/01/16/how-top-brands-nurture-their-online-communities-04835959 (accessed on 30 July 2018).

Gower, Geoff. 'Marketing to Millennials: The Rise of Content Co-creation'. *The Guardian*, 3 November 2013. Available at: https://www.theguardian.com/media-network/2014/nov/03/marketing-millennials-content-creation   (accessed on 19 July 2018).

Hill, Janelle B. 'Leading Through Digital Disruption'. Gartner, 2017. Available at: https://www.gartner.com/imagesrv/books/digital-disruption/pdf/digital_disruption_ebook.pdf (accessed on 30 July 2018).

Hollenbeck, Candice R., and Andrew M. Kaikat. 'Consumers' Use of Brands to Reflect Their Actual and Ideal Selves on Facebook'. *International Journal of Research in Marketing* (16 September 2012). Available at: https://media.terry.uga.edu/socrates/contact/documents/2016/02/29/Consumers_use_of_brands_to_reflect_their_actual_and_ideal_selves_on_Facebook.pdf (accessed on 27 July 2018).

Howarth, Brad. 'What Zappos Is Doing to Personalise Customer Experience and Services'. CMO, 25 May 2018. Available at: https://www.cmo.com.au/article/641587/what-zappos-doing-personalise-customer-experience-services/ (accessed   on   11 September 2018).

Hughes, Melissa. 'Soraya Mehdizadeh, Undergraduate Psychology Student, Finds Facebook Fiends Tend to be Narcissistic and Insecure'. York University, 7 September 2010. Available at: http://research.news.yorku.ca/2010/09/07/soraya-mehdizadeh-york-university-undergraduate-student-finds-facebook-fiends-tend-to-be-narcissistic-and-insecure/ (accessed on 19 July 2018).

IBM. 'IBM Survey Shows Strengths, Gaps in U.S. Office Buildings'. IBM, 29 April 2010. Available at: https://www-03.ibm.com/press/us/en/pressrelease/30191.wss (accessed on 30 July 2018).

IDG. 'Understand How Organizations Evolve to a Digital Business Model'. Digital Business Survey, 2018. Available at: http://resources.idg.com/download/white-paper/2018-digital-business (accessed on 30 July 2018).

IDG, and Dell. 'Digital Transformation: Crossing the Chasm from IT to the Business'. White Paper, 2017. Available at: http://marketing.dell.com/Global/FileLib/eLearning/Digital-Transformation-Crossing-the-Chasm.pdf (accessed on 30 July 2018).

Igniyte. *The Business of Reviews Infographic*. 24 April 2015. Available at: https://www.igniyte.co.uk/reports/the-business-of-reviews-infographic/ (accessed on 19 July 2018).

Integer, and Coca-Cola Retailing Research Council. 'Untangling the Social Web: Insights for Users, Brands, and Retailers'. Coca-Cola Retailing Research Council, March 2012. Available at: http://www.ccrrc.org/wp-content/uploads/sites/24/2014/02/Untangling-the-Social-Web_Part-31.pdf (accessed on 27 July 2018).

Jenner, Sarah, and Daniel Suss. 'Socialization as Media Effect'. John Wiley & Sons, 2017. Available at: http://sarah.genner.cc/uploads/5/1/4/1/51412037/socialization_as_media_effect.pdf (accessed on 27 July 2018).

Kaplan, Melanie D. G. 'Intelligent Elevators Answer Vertical Challenges'. ZDNet, 17 July 2012. Available at: https://www.zdnet.com/article/intelligent-elevators-answer-vertical-challenges/ (accessed on 30 July 2018).

Kapoor, Gaurav. 'Doctors and Digital Devices: India Bridging the Gap'. NASSCOM, 16 February 2016. Available at: http://blogs.nasscom.in/doctors-and-digital-devices-india-bridging-the-gap/ (accessed on 30 July 2018).

Kemp, Simon. 'Digital in 2018: World's Internet Users Pass the 4 Billion Mark'. We Are Social, 30 January 2018. Available at: https://wearesocial.com/blog/2018/01/global-digital-report-2018 (accessed on 30 July 2018).

Kennedy, Gemma, and Elvira Bolat. 'Meet the HENRYs: A Hybrid Focus Group Study of Conspicuous Luxury Consumption in the Social Media Context—Competitive Paper'. 2017. Available at: http://eprints.bournemouth.ac.uk/29423/3/AM17_0335_competitive.pdf (accessed on 27 July 2018).

Khan, Danish. 'Xiaomi Widens Smartphone Market Share Gap with Samsung in India: Canalys'. *ET Telecom*, 24 April 2018. Available at: https://telecom.economictimes.indiatimes.com/news/xiaomi-widens-smartphone-market-share-gap-with-samsung-in-india-canalys/63891082 (accessed on 26 July 2018).

Kline, David. 'Behind the Rise and Fall of China's Xiaomi'. *Wired*, 22 December 2017. Available at: https://www.wired.com/story/behind-the-fall-and-rise-of-china-xiaomi/ (accessed on 26 July 2018).

Kornwitz, Jason. 'When It Comes to Social Media, Consumer Trust Each Other, Not Big Brands'. News@Northeastern, 18 September 2017. Available at: https://

news.northeastern.edu/2017/09/18/when-it-comes-to-social-media-consumers-trust-each-other-not-big-brands/?_ga=2.129642700.1593677796.1528708376-336971180.1528708376 (accessed on 26 July 2018).

Krasnova, Hanna, Sarah Spiekermann, Ksenia Koroleva, and Thomas Hildebrand. 'Online Social Networks: Why We Disclose'. *Journal of Information Technology* (June 2010). Available at: https://www.researchgate.net/publication/220220751_Online_Social_Networks_Why_We_Disclose (accessed on 30 July 2018).

Lindgren, Jennifer. 'Deep Ellum Film Processing Business Is Surviving Digital Age', CBS, 4 June 2018. Available at: http://dfw.cbslocal.com/2018/06/04/film-processing-business-surviving-digital-age/ (accessed on 11 September 2018).

Mahindra Rise. 'Spark the Rise with Mahindra'. Mahindra, 11 April 2011. Available at: http://www.mahindra.com/news-room/press-release/1313067196 (accessed on 19 July 2018).

Mathur, Vishal. 'Why Does a Phone Go Out of Stock Within Seconds of Going on Sale?' Livemint, 6 September 2014. Available at: https://www.livemint.com/Industry/N9u9LgpBPphua8nejlNqCK/Why-does-a-phone-go-out-of-stock-with in-seconds-of-going-on.html (accessed on 26 July 2018).

McAteer, Oliver. 'Does A.R.'s Sneaker Play Spell the End of Brick and Mortar?' Campaign Live, 11 June 2018. Available at: https://www.campaignlive.com/article/does-ars-sneaker-play-spell-end-brick-mortar/1484666 (accessed on 11 September 2018).

McCarthy, John. 'How Social Media Told Protein World Where to Stick Its #beachbodyready Ads'. The Drum, 29 April 2015. Available at: http://www.thedrum.com/news/2015/04/29/how-social-media-told-protein-world-where-stick-its-beachbodyready-ads (accessed on 30 July 2018).

McGrath, Siobhan. 'The Impact of New Media Technologies on Social Interaction in the Household'. National University of Ireland, 19 April 2012. Available at: https://www.maynoothuniversity.ie/sites/default/files/assets/document/SiobhanMcGrath.pdf (accessed on 27 July 2018).

Mukherjee, Writankar. 'Around 40% Indians Want to Change Mobile Phones Within a Year: Study'. *The Economic Times*, 9 March 2018. Available at: https://economictimes.indiatimes.com/tech/hardware/around-40-indians-want-to-change-mobile-phones-within-a-year-study/articleshow/63238890.cms (accessed on 26 July 2018).

Munro, Kali. 'Conflict in Cyberspace: How to Resolve Conflict Online'. Kalimunro.com, 2002. Available at: http://kalimunro.com/wp/articles-info/relationships/article (accessed on 30 July 2018).

Nair, Bhagyashree, and Richa Maheshwari. 'Demonetisation-hit Luxury Home Decor Business Rebounds to New Highs'. *The Economic Times*, 29 April 2017. Available at: https://economictimes.indiatimes.com/magazines/panache/

demonetisation-hit-luxury-home-decor-business-rebounds-to-new-highs/
articleshow/58424997.cms (accessed on 27 July 2018).

Nayar, Aashimita. 'How Dating Apps Are Radically Changing the Rules of Romance
in India'. *Huffington Post*, 15 June 2016. Available at: https://www.huffington
post.in/2015/02/05/apps-websites-digital-dat_n_6415996.html (accessed on 11
September 2018).

Nielsen. Ecommerce: Evolution or Revolution in the Fast-moving Consumer
Goods World. Nielsen Global E-commerce Report, August 2014. Available at: http://
www.nielsen.com/content/dam/nielsenglobal/apac/docs/reports/2014/
Nielsen-Global-E-commerce-Report-August-2014.pdf (accessed on 17 July 2018).

Olenski, Steve. '7 Marketing Lessons from the ALS Ice Bucket Challenge'. *Forbes*,
22 August 2014. Available at: https://www.forbes.com/forbes/welcome/?to
URL=https://www.forbes.com/sites/steveolenski/2014/08/22/7-marketing-
lessons-from-the-als-ice-bucket-challenge/&refURL=https://www.google.co.
in/&referrer=https://www.google.co.in/ (accessed on 31 July 2018).

Para, Jen. 'HBJ Announces Rankings of Houston's Healthiest Employers'. *Houston
Business Journal* (17 May 2018). Available at: https://www.bizjournals.com/
houston/news/2018/05/17/hbj-announces-rankings-of-houston-s-healthiest.
html?ana=yahoo&yptr=yahoo (accessed on 30 July 2018).

Pew Internet. 'Substantial "Reciprocity" Across Major Social Media Platforms'.
Pew Research Center, 27 February 2018. Available at: http://www.pewinternet.
org/2018/03/01/social-media-use-in-2018/pi_2018-03-01_social-media_0-04/
(accessed on 27 July 2018).

Phelon, Promise. 'Why Consumers Don't Trust Your Brand Content and How to
Fix It'. *Forbes*, 24 January 2017. Available at: https://www.forbes.com/forbes/
welcome/?toURL=https://www.forbes.com/sites/promisephelon/2017/01/24/
why-consumers-dont-trust-your-brand-content-and-how-to-fix-that/&ref
URL=https://www.google.co.in/&referrer=https://www.google.co.in/ (accessed
on 31 July 2018).

PowerReviews, and Northwestern University. 'From Reviews to Revenue—
Volume 1: How Star Ratings and Review Content Influence Purchase'. Power
Reviews, 2016. Available at: https://www.powerreviews.com/wp-content/uploads/
2016/04/Northwestern-Vol1.pdf (accessed on 25 July 2018).

PR Newswire. 'New Study Shows User-generated Content Tops Marketing Tactics
by Influencing 90 Percent of Shoppers' Purchasing Decisions'. PR Newswire, 19 June
2017. Available at: https://www.prnewswire.com/news-releases/new-study-shows-
user-generated-content-tops-marketing-tactics-by-influencing-90-percent-of-
shoppers-purchasing-decisions-300475348.html (accessed on 31 July 2018).

Prensky, Marc. 'Digital Natives, Digital Immigrants'. In *On the Horizon*. MCB
University Press, October 2001. Available at: https://www.marcprensky.com/

writing/Prensky%20-%20Digital%20Natives,%20Digital%20Immigrants%20
-%20Part1.pdf (accessed on 27 July 2018).

Saurabh, Saket. 'Millennials & New Media Platforms: A Match Made in Heaven'.
NextBigWhat, 27 April 2017. Available at: https://www.nextbigwhat.com/india-
millennials-media-platforms-297/ (accessed on 18 July 2018).

Shaojung Sharon Wang, and Michael A. Stefanone. 'Showing Off? Human Mobility
and the Interplay of Traits, Self-disclosure, and Facebook Check-ins'. *Social
Science Computer Review* 31, no. 4 (2013). Available at: http://citeseerx.ist.psu.
edu/viewdoc/download?doi=10.1.1.723.4896&rep=rep1&type=pdf (accessed on
27 July 2018).

Sharma, Manoj. 'Online Gallery: Internet Gives Artists a New Marketplace'.
Hindustan Times, 17 September 2016. Available at: https://www.hindustantimes.
com/delhi-news/online-gallery-internet-gives-artists-a-a-new-marketplace/
story-UYutypVeDpg1bnboQDVkxM.html (accessed on 17 July 2018).

Shrestha, Khusbu. '50 Important Stats You Need to Know About Online Reviews'.
Vendasta Blog, 21 May 2018. Available at: https://www.vendasta.com/blog/50-
stats-you-need-to-know-about-online-reviews (accessed on 17 July 2018).

Singhi, Abheek, and Nimisha Jain. 'The Rise of India's Neo Middle Class'. *HT
Mint*, 4 October 2016. Available at: http://www.livemint.com/Politics/HY9TzjQzlj
CZNRHb2ejC2H/The-rise-of-Indias-neo-middle-class.html (accessed on 17 July
2018).

Smith, Aaron, and Monica Anderson. '2. Online Reviews'. Pew Research Center,
19 December 2016. Available at: http://www.pewinternet.org/2016/12/19/online-
reviews/ (accessed on 25 July 2018).

SocioAdvocacy. 'Mind-blowing Stats about Indians on Instagram'. SocioAdvocacy,
2 May 2017. Available at: https://www.socioadvocacy.com/blogs/instagram-
audience-in-india-2017/ (accessed on 27 July 2018).

Spangler, Todd. 'Younger Viewers Watch 2.5 Times More Internet Video Than TV
(Study)'. *Variety*, 29 March 2016. Available at: http://variety.com/2016/digital/
news/millennial-gen-z-youtube-netflix-video-social-tv-study-1201740829/
(accessed on 18 July 2018).

———. 'Americans Are Watching Less Traditional TV as Smartphone Media
Usage Booms'. *Variety*, 27 June 2016. Available at: http://variety.com/2016/digital/
news/live-tv-declining-smartphone-boom-nielsen-1201804202/ (accessed on
18 July 2018).

Suler, John. 'The Online Disinhibition Effect'. *Cyberpsychology & Behaviour*
(2004). Available at: https://pdfs.semanticscholar.org/c70a/ae3be9d370ca1520db
5edb2b326e3c2f91b0.pdf (accessed on 30 July 2018).

Tassi, Paul. 'Wii U Sales Down 36%, Nintendo Executives Taking Pay Cut'. *Forbes*, 29 January 2014. Available at: https://www.forbes.com/sites/insertcoin/2014/01/29/wii-u-sales-down-36-nintendo-executives-taking-pay-cut/#47772299cf31 (accessed on 25 July 2018).

Terdiman, Daniel. 'Here's How People Say Google Home and Alexa Impact Their Lives'. Fast Company, 1 May 2018. Available at: https://www.fastcompany.com/40513721/heres-how-people-say-google-home-and-alexa-impact-their-lives (accessed on 27 July 2018).

Wakeman, Gregory. 'What Studio Executives Think of Rotten Tomatoes'. Cinema Blend, 2016. Available at: http://www.cinemablend.com/new/What-Studio-Executives-Think-Rotten-Tomatoes-80747.html (accessed on 18 July 2018).

Wave8. 'The Language of Content'. UM Wave, 2017. Available at: https://wave.umww.com/assets/pdf/wave_8-the-language-of-content.pdf (accessed on 19 July 2018).

WE Worldwide, and Georgetown University. 'Digital Persuasion: How Social Media Is Being Used to Influence Perceptions, Actions & Support for Causes'. 2013. Available at: http://csic.georgetown.edu/wp-content/uploads/2016/12/digital-persuasion.pdf (accessed on 11 September 2018).

West, Matt. '#PRfail: Twitter Users Give British Gas a Roasting as It Hosts Online Q&A on Social Media Site on Same Day It Hikes Bills'. This Is Money, 17 October 2013. Available at: http://www.thisismoney.co.uk/money/bills/article-2465053/PRfail-Twitter-users-British-Gas-roasting-hosts-online-Q-amp-A-social-media-site-day-hikes-prices-9-2.html (accessed on 30 July 2018).

World Economic Forum. *Digital Media and Society: Implications in a Hyperconnected Era*. World Economic Forum, January 2016. Available at: http://www3.weforum.org/docs/WEFUSA_DigitalMediaAndSociety_Report2016.pdf (accessed on 19 July 2018).

About the Author

Ray Titus is Professor of Marketing and Dean at the Alliance School of Business, Alliance University, Bengaluru. Recognized as a 'consumer lifestyle expert' by the *Economic Times*, Ray is sought after by the industry and academia for his marketing and consumer behaviour expertise. His articles and expert comments have featured in the *Economic Times*, Business Insider, the *Globe and Mail*, the *Hindu Business Line* and other leading dailies. Ray was recently listed among the '10 Indian Business Bloggers You Need to Follow in 2018' for his blog 'Buyer Behaviour'. He is also the author of the business bestseller *Yuva India: Consumption and Lifestyle Choices of a Young India*. He makes his home in Bengaluru with wife Alphy, son Jaden and daughter Brooklyn.

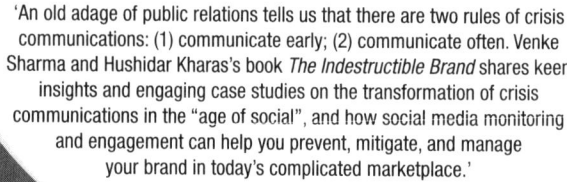

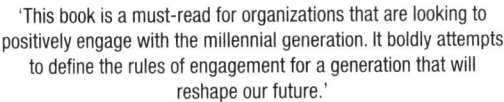